ISLAM & NATURAL PHILOSOPHY

First published in the UK by Beacon Books and Media Ltd
Earl Business Centre, Dowry Street, Oldham, OL8 2PF, UK.

www.beaconbooks.net

ISBN 978-1-915025-55-5 Paperback
ISBN 978-1-915025-56-2 Hardback
ISBN 978-1-915025-57-9 Ebook

Cataloging-in-Publication record for this book is available from the British Library.

Cover design by Raees Mahmood Khan
Cover image by Evgeni Tcherkasski on Unsplash

ISLAM & NATURAL PHILOSOPHY

PRINCIPLES OF DAQIQ AL-KALAM

BASIL ALTAIE

BEACON BOOKS

To My Wife, Nada

Contents

Acknowledgements

I would like to thank Dr. Abdal Hakim Murad (Tim Winter) from the Divinity Department of the University of Cambridge, U.K. for encouraging me to publish this book in English.

My gratitude goes to Professor Ilyas Celebi and Dr. Mehmet Bulgen from the *Kalām* Department at the University of Marmara, Turkey, for sharing with me useful discussions about topics in *kalām*, and also adopting this book as a textbook for their postgraduate course.

I would further like to thank *Kalām* Research and Media (UAE) for partially funding the translation, and for editing this book.

Thanks also to Dr. Zarina Yousuf-Bonass for editing this manuscript and to my publisher, Mr. Jamil Chishti, and his team for the care they have provided.

Finally, my deep thanks go to my wife, Nada, for the great support and sacrifices she has made during the many years it has taken me to write my book.

Note about Transliteration and Use of Terms

I have adopted the transcription of the Arabic words rather than their transliteration. This should enable the reader to get the vowels of the word, as much as possible, nearer to the Arabic sound. The reader should then be able to differentiate between the singular and the plural, for example, *Jawhar* is singular whereas *Jawāhir* is plural. Also, *ʿaraḍ* is singular and *aʿrāḍ* is plural. Also of interest to note, is that Ḥayyīz is different from *Taḥayyuz*, and so on. Therefore, some caution should be taken in some places where a differentiation might be necessary.

Preamble

Kalām is one of the basic pillars of Islamic thought. Begun during the early decades of the 2nd century after Hijra (9th century A.D.), this trend aimed to resolve some questions in the Islamic creed by adopting a rational interpretation of religious teachings. Soon the teachers of *Kalām* (the Mutakallimūn) found it necessary to get involved in discussing questions related to topics on natural philosophy, such as space, time, matter, and their constituent. They tried to construct a system of thought and worldview that comply with the teachings of the Qur'an. In this book, I present this part of their thought, which concerns their views of the natural world formulated in a modern context of expressions. Such a formulation is necessary to understand the true meaning and aims of *kalām* philosophy.

My first encounter with *kalām* was about thirty years ago, when I became interested in this area of Islamic thought. On reading Shlomo Pines' book, *Beiträge zur islamischen Atomenlehre* (*Studies in Islamic Atomism*), which was translated from German into Arabic, I realised that this neglected area of Islamic thought contains important connections with many of the concepts and views in natural philosophy.

To get a better understanding, I visited the University of Mosul library, looking for books and manuscripts containing original contributions on *kalām*. I was delighted to see that most of the references mentioned in Pines' book were still available. I found the books of al-Ashʿari, al-Bāqillānī, al-Jūwaynī, al-Ghazālī, Ibn Ḥazm, Ibn Rushd, al-Khayyāt, as well as other books which were discovered at a later time. These include the encyclopeadia of al-Qāḍī ʿAbd al-Jabbār, one of the prominent leaders and historians of the Muʿtazilis (*al-Mughnī fi Abwāb al-ʿadl wa al-Tawḥīd*), as well as the books of al-Naysābūrī and Ibn Matawayh, along with many others. For me, this was akin to discovering a gold mine. Since then, I have become a permanent resident in that section of the library!

As I mined through those books over four years, I recognised that there were two topics which the Mutakallimūn were concerned with; the first of these was the part concerning theological questions, which they called *Jalīl al-kalām,* and the second part concerned the physical questions relating to natural philosophy, which they called *Daqīq al-kalām.* Being a trained theoretical physicist, I was able to grasp much of the content connected to natural philosophy, while discarding the part regarding detailed theological questions (which were mostly on metaphysical enquiries in Islamic creed). On certain issues, however, it was hard to dissociate the theological questions from views which fell under natural philosophy. However, I was pleased that some of the early leaders of the Mutakallimūn devoted a section of their books to *Daqīq al-kalām,* and discussed natural philosophy, while devoting other sections to theological problems i.e. *Jalīl al-kalām* (al-Ashʿari's book, *Maqalāt al-Islamyyīn,* for example). This distinction was necessary at this time of thought development in order to identify the influence of belief when it came to constructing a worldview.

On studying the topics of *Daqīq al-kalām,* I was always interested in following the motivation behind the views of those Mutakallimūn. For this reason, I had to go through detailed analysis and discussion of topics in Islamic theology, such as Divine action and understanding the process of change in nature, according to the views of the Mutakallimūn. There I had to consider for example, the question of causality in nature and human behaviour, including the issues related to Divine will, predestination and freewill. In this respect, I had to confine my study and concentrate on the physical issues related to the subject.

An important part of my scientific enquiry, when studying diverse issues, was to recognise basic elements and principles, as well as the embodied assumptions that served to construct the Mutakallimūn's views on those issues. Such an identification helped me understand the pillars and structure of their thought. Thus, after thoroughly studying the main topics of *Daqīq al-kalām,* I was able to identify some basic principles, propositions and doctrines that formed the Mutakallimūn's views on natural philosophy. It is these that I call *The Principles of Daqīq al-Kalām,* and which are the subject of study, analysis, and discussion in this book.

It should be noted that my presentation of the Mutakallimūn's thoughts and principles on natural philosophy is not intended to be historical, nor is it meant to be put in a context of comparative studies. Instead, it is an analysis of their philosophical and scientific understanding of the natural world. This might require, in some places, making a comparison with rival notions in Greek philosophy so as to identify differences, but certainly this comparison is not exhaustive. In this respect, I have taken care to identify the true meanings for some expressions used by the Mutakallimūn, such as *jawhar*, and *'araḍ*, which have corresponding terms in the Greek philosophy of substance and accident. The intention of this is to bring the expressions to their true meaning in Arabic, as expressed by the Mutakallimūn.

In his book, *The Philosophy of the Kalām*,[1] Harry Wolfson, of Harvard University, followed the methodology of conjecture and verification without having a frame of reference for his ideas when probing the thoughts of the Mutakallimūn. Such an approach caused Wolfson's analysis to lean on assumptions which were very much influenced by his personal attitude. In this study, I analyse the living texts of *kalām* by adopting two basic references: namely the Arabic, in which the authentic text is written, and those that have been verified by science.

During my journey investigating *kalām*, I visited the Faculty of Arts at my university and met with Professors of History and Philosophy, seeking to discuss the legacy of *kalām* with them. I was disappointed to learn that *kalām* was not part of their curricula. Perhaps this was because they did not have a faculty member who was specialised enough to teach this subject. Nevertheless, it was a kind gesture from the Dean of Faculty to invite me to teach a course on "Topics in *Kalām* and Philosophy." Since then, I have learned a lot and was able to publish my first article, *The Scientific Value of Daqīq al-Kalām,* in 1994.

My own area of specialisation in Quantum Field Theory and the Theory of General Relativity was a great help to me as it enabled me to recognise, not only the theoretical roots, but also the scientific and philosophical applications of the principles of *Daqīq al-kalām*. My encounters with the philosophical implications of quantum mechanics allowed me to appreciate areas where these principles could be applied, either to gain a better understanding or to provide clearer explanations when I found a fruitful interplay between *Daqīq al-kalām* and contemporary philosophy of science. This was elaborated over many years of struggle through discussions and involvements in several debates concerning science and religion. During this period, I attended and contributed to several conferences held at Oxford University and other British universities, which were organised by the British Science and Religion Forum. I learned a great deal and found that Islamic *kalām* had much to say on those discussions. Similarly, I have also found that some questions and arguments in current science and religion debates were also hot topics of the old *kalām*, with some contemporary views echoing the old ideas of the Mutakallimūn.

---

1 Wolfson, H. A. (1976). *The Philosophy of the Kalām*. Cambridge, MA: Harvard University Press.

All this provided me with the necessary knowledge to write this book, which was first written in Arabic under the title, *Daqīq al-Kalām: the Islamic Approach to Natural Philosophy*, and which was published in 2010. In it, I presented the main principles of *Daqīq al-kalām* and discussed some problems of current interest. The book was published as a second, expanded, edition by *Kalām* and Research Media (KRM) in 2018.

I find that a new formulation of *kalām*, using the basic doctrines of the early works of the Mutakallimūn could lay the foundation for a major transformation in Islamic thought. The old *kalām* is no longer useful; it is simply a monumental legacy that may, in their historical context, serve to teach us some useful lessons. However, we can re-build *kalām* and set it as a new venture, free from those enigmas which led theological *kalām* (*Jalīl al-Kalām*) to descend into a vicious debate. This reformation is very important; first, for Muslims, to enable them to break away from those historical and cultural boundaries and limitations which have captivated their minds for centuries. Secondly, the revival of *kalām* is important for the rest of the world, which is suffering from the riddle of atheism-theism and the puzzle of scientism versus belief. Furthermore, there are many fundamental questions which contemporary philosophers of science are grappling with in vain. It may be that some of these questions could gain enlightenment from the new *kalām*.

Part I
Historical Background

Chapter One

Introduction

A SHORT HISTORY

This book aims to uncover the basic tenets in, what was, an important chapter in the history of Islamic thought; one which was presented as rational discourse concerned with understanding the natural world and its phenomena. This area of Islamic thought was part of a system of thought, which grew up as a rival of philosophy and was named *kalām*.

Generally, *kalām* is concerned with providing a rational Islamic worldview on all matters regarding religious belief, as well as the social and personal aspects of life. It was aimed at laying the foundations for the Sharīʿa (religious) laws. The part concerned with questions of natural philosophy is called *Daqīq al-Kalām* (fine *kalām*) while the other, which deals with theological questions is called *Jalīl al-kalām* (the major or coarse *kalām*). These parts were not always set out in separate or independent contexts. In some cases, both sections were interwoven, especially when dealing with theological questions. However, for the sake of clarity, and since I am hoping to re-construct the *kalām* system of thought in a modern context, I selected the topics concerning natural philosophy and regarded them as being distinct in themselves. This certainly enabled me to analyse the approach undertaken by the Mutakallimūn and disseminate the results in a clearer and more efficient way.

The subject of *kalām* was developed during the first few decades of the 8th century A.D, two hundred years after the Prophet Muhammad (peace be upon him) died, as a rational approach for elaborating the truth of Islam as revealed in the Qur'an and the teachings of the Prophet. It was an intellectual venture aimed at expressing an Islamic worldview in terms of the natural world and human beings. Additionally, it welcomed input from all Muslim intellectuals, thereby inspiring a burst of contributions. These generated several tenets and schools of thought; one of the earliest of which was called the *Mu'tazila*. Scholars of this school tried to establish the basis for the Islamic fiqh (jurisprudence) by which Sharīʿa law could be deduced through the rational understanding of the Qur'an and the teachings of the Prophet. Initially, the venture went well and achieved great success through Abū Ḥanīfa al-Nuʿmān (699–767) and his school in Basra, the Malikī school, founded by Malik Ibn Anas (711–795), the Sharīʿa school of fiqh, founded by al-Shafiʿī (767–820), and the school of Ahmad Ibn Ḥanbal (780–855), which was set up later. The difference between these schools of fiqh lay in the principles they upheld and the resources they used for deducing Sharīʿa laws. These set of rules are called *Usūl al-Fiqh*.

Abū Ḥanīfa relied on a rational analysis of problems with some support from the Qur'an and the narrations of the Hadith of the Prophet. The Malikī and Shāfʿī schools took a moderate position and relied on rational analysis as much as authentic narrations from the Ḥadīth to deduce Sharīʿa laws. The Ḥanbalī school was almost fixated on following the Ḥadīth (narrations) of the Prophet and the views of his companions (Saḥāba), a trend that established a more dogmatic approach to Sharīʿa.

An unfortunate incidence during the early history of Islamic thought occurred when the Muʿtazilis became engaged in a conflict regarding dogma, especially the Ḥanbalī school. The main point of the conflict was the issue known as the *Problem of the Creation of the Qur'an*. Here, the dominant leaders of the Muʿtazilis, supported by the state, insisted that all the clergy should adopt the notion that the Qur'an was created (*makhlūq*). This approach, they felt, would allow them to consider the Qur'an and its content from both the rational as well as the historical viewpoint. The Ḥanbalī school, led by Ahmad Ibn Ḥanbal, opposed this ruling and fell into a dispute with the Abbasid regime of al-Muʿtasim (794–842), which was overrun with Muʿtazilis. As a consequence of this disagreement, Ibn Ḥanbal was jailed and tortured, and his followers prosecuted by the Muʿtazilis.

At the beginning of the 10th century, Abū al-Ḥasan al-Ashʿari, a descendant of the Muʿtazilis, formed a new school of *Kalām*, called the *Ashʿaris*. Greatly influenced by the suffering and agony of the past conflict between the Muʿtazilis and the Ḥanbalis, many people subscribed to this new school. Ever since then, this school has flourished.

The Ashʿari school established different views of *kalām* and attempted to drastically move away from the Muʿtazili school on questions related to the divinity, divine attributes and seeing the divine on doomsday, amongst other issues. The decades that followed witnessed conflicts between Ashʿaris and Muʿtazilis, the victors being the one which gained support from the ruler of the time. This situation continued until al-Qādir bi'llāh came to power (991–1031). During his reign, this Abbasid Caliph issued the famous al-Qādir declaration by which he banned all *kalām* schools and endorsed the Ḥanbalī school in the Islamic system of belief. Sharīʿa laws were assigned to jurisprudent scholars to deal with according to their particular school of study and deduction.

Here, we should note that the *kalām* schools were not religious schools. Although, historically, they were considered theologians, the Mutakallimūn (practitioners of *kalām*) were innovative scholars who were not necessarily theologians. The four main sunni schools: the Ḥanafī, Mālikī, Shāfiʿī and Ḥanbalī, however, are all schools of theology and Sharīʿa. Thus, when *kalām* was banned, the only venue for meaningful discussion and studies were these dogmatic schools of theology and Sharīʿa. Consequently, the free, intellectual atmosphere declined, and very little innovative or intellectual material in *kalām* was produced thereafter.

As mentioned above, *kalām* was intended to establish a rational approach to understanding God, humankind, and nature according to the Islamic system of belief. *Kalām* considered two types of questions: those about divinity, where matters directly related to belief (including divine attributes, resurrection of the dead, seeing God, and punishment and reward in the afterlife), and those questions related to divine knowledge, will, and power. These queries, in turn, led to the question of human free will, as advocated by the Muʿtazilis school of *Kalām*, and the counter proposal of self-acquisition (*kasb*) of actions, which was held by the Ashʿaris school. This has led to discussions on the concept of space, time, motion, force, and many other aspects of the physical world. Thus, this part might also be called the physical theory of *kalām*.

AIMS OF THE BOOK

In this book I am going to discuss the basic tenets of Islamic approach to natural philosophy. I have articulated these as the '*principles of Daqīq al-Kalām*', having authenticated them from the original sources of *kalām* and analysed them from a modern perspective. I am seeking to present the tenets in a modern philosophical and scientific context, as many of the principles and doctrines are compatible with theories in modern-day physics. Indeed, it would be interesting to see how far these concepts of *Daqīq al-kalām* can be taken along this line.

It might be noted that my designation of the basic view of *kalām* concerning the physical issues that I deal with here, is somewhat selective, since my aim is to construct a basic view based on the legacy of *kalām* and to reconstruct it from a new perspective on natural philosophy. One of the main objectives of this book is to re-evaluate some common impressions and pre-conceptions that have been established by contemporary authors. Several of these perceptions were generated through misreading the *kalām* texts or reading them from secondary sources which were either written by opponents of *kalām*, or the opponents of some schools of *kalām*. For example, if we take the words of ʿAbdul Qāhir al-Baghdādī, the famous Ashʿari, about the Muʿtazilis literally, we would get the wrong impression about them, especially his views about Ibrāhīm al-Naẓẓām, where he calls those views of al-Naẓẓām scandalous!

Some of the errors in understanding certain concepts are due to the incorrect translation of several terms. This, we can see clearly in a number of contemporary studies on various topics of the physical theory of *kalām*. Some concepts have not received the proper appreciation. This may be due to the authors and scholars, who dealt with them, searching for their historical origin, and believing that they must have a Greek or Indian source. In this case, they would have overlooked the fact that the cultural wealth of those who initiated *kalām* was so deeply rooted in the history of thought as to boost their intellectual capabilities to high levels. This privilege enabled them to deal with delicate matters which needed an intense theoretical involvement. Accordingly, the results obtained by such scholars has been mostly inconclusive. In this respect, I point to the studies of Shlomo Pines (*Studies in Islamic Atomism*) and Harry Wolfson (*The Philosophy of the Kalām*).

There are erroneous myths that have been widely circulated about some schools of *kalām*. One concerns the position of the Muʿtazilis on causality. In this book I show that the understanding of natural causality of the Muʿtazilis is not much different from that of the Ashʿaris; if we take the views of Qaḍī ʿAbd al-Jabbār to represent the first school and those of al-Bāqillānī to exemplify the other. We see that both scholars adopted common principles according to which they formed their common views. Within the same myth lies the claim that the Muʿtazilis understood the world to be deterministic, whereas the Ashʿaris assumed it was occasional. The investigation presented in this book shows that both the Muʿtazilis and the Ashʿaris believed in the same principles and adopted similar views as far as the natural world was concerned. The actual difference between them was in their views about divinity and subjects within *Jalīl al-kalām*. This ability to distinguish between similarities and differences is one of the fundamental benefits of studying the legacy of *kalām* under the separate classifications of *Daqīq* and *Jalīl al-kalām*.

REFORMING ISLAMIC THOUGHT

The Muslim community is in urgent need of reforming their way of thinking if it is to remove itself from the stagnation they have suffered during the last three centuries. Two factors play an essential role in any successful transformation of Islamic thought. Firstly, the change should come from within the Islamic way of thinking itself and, secondly, it should have a strong philosophical basis, stemming from original Islamic sources which are compliant with the modern state of knowledge and life. The previous attempts to revitalise *Kalām*, which were undertaken during the last decades of the 19th century and the beginning of the 20th century, have been reviewed by Özervarli.[2]

My proposal for effecting a transformation in Islamic society is based on revitalising *kalām* to become an efficient methodology for analysis and deduction in religion and modern Islamic thought. This can be done in two stages. First, we should revitalise *Daqīq al-kalām* such that this tradition of thought stands for the Islamic view of nature and science. The fact that the principles of *Daqīq al-kalām* conform with concepts of modern physics and cosmology would enable us to achieve such a goal without much trouble. Second, once the principles and methodology of *Daqīq al-kalām* are established, some essential problems in natural sciences, social sciences, religion, and the arts will need to be re-analysed and studied according to the new methodology. Questions, such as the epistemological value of science, determinism and causality in the natural world, biological evolution, the design argument, and others, might be discussed and analysed within the context of neo-*kalām* interfacing with scientific facts, thereby developing a worldview that shares the achievements of modern science. This might be initiated by some sample case studies to demonstrate the effectiveness and efficiency of the new methodology and principles.

Once *Daqīq al-kalām* is set in scholastic studies, other questions, such as divine action, consciousness, free will and predestination can be studied on the grounds provided by the earlier studies of *Daqīq al-kalām*. No doubt these studies will also serve as an essential step in delving into *Jalīl al-kalām* as well, which is a much more subtle topic matter. With new light shed on the topics of *Jalīl al-kalām*, matters of Sharīʿa law can then be discussed and the proper rules applied.

I believe that if one can develop a trend of change that motivates critical thinking based on proper foundations, and proposes an efficient scheme to deal with revising Islamic thought, including Islamic Sharīʿa, then a real

2 Özervarlı, M. Sait. "Attempts to Revitalize *Kalām* in the Late 19th and Early 20th Centuries." *The Muslim World* 89, no. 1 (1999): 90–105.

transformation might be initiated. To be successful, this transformation should stem from the original sources of Islam, thereby preserving the mission of Islam and developing it into a workable modern system of thought.

The decision for a need to revitalise *Kalām* raised a number of vital issues that had to be considered. First, we should take into account that *kalām* is traditionally not favoured and is a subject in disrepute. Most traditional theologians of the 14th century, and later, considered discourse on *kalām* to be a distortion of Islam and a potential danger that might cause a pious Muslim to be dislodged from their religion. This attitude is also widespread amongst Muslims today, and I confess to having faced difficulties in discussing *Daqīq al-kalām* with scholars who were brought up according to traditional (Salaf) beliefs. Second, we should remember that the practical failure of scholars' efforts during the 19th and early 20th centuries was partially due to the fact that they were not acquainted with modern scientific knowledge and, consequently, they could not harmonise the truth of Islam alongside that of science. Muhammad ʿAbduh in Egypt and Shiblī Nuʿmanī in India, are examples of such scholars. Certain facts on quantum physics, relativity and modern biology were unknown to those scholars and hence they applied the concepts of classical physics and their limited knowledge of other natural sciences. Subsequently, ʿAbduh was unable to practically implement his claims of harmony between science and Islam, while Shiblī Nuʿmānī could not adapt the concepts of the mechanistic world of classical physics and assume deterministic causality, which is in contradiction with the ideas of *kalām*.

The new approach suggested for the transformation of Islamic thought opens the way for a well-founded methodology of theological development (*Ijtihād*); discussions around this fundamental trend of jurisprudence have been threatened with being banned. This trend concerns the sources for establishing Sharīʿa, the rules for accepting and rejecting narrations of hadīth, interpreting verses of the Qurʾan dealing with Sharīʿa, and deducing its laws. This is, potentially, a pivotal point when it comes to establishing new trends in Islamic jurisprudence.

Some modern thinkers and reformists have been captivated by the methodology of the traditional theologians (the *Salaf)* and their understanding of various terms in the Qurʾan and the Ḥadīth, including mistakes, both deliberate and accidental, that they made in those interpretations. However, I feel that we should neither allow ourselves to become involved in such captivation, nor accept their methodology in full. Instead, we should allow for whatever we find truthful to be in our hands. For this purpose, we should be able to provide strong arguments from the Qurʾan, the Arabic language, and our intellect, using a common ground of agreement, which supports the fact that our approach maintains the originality of Islamic teachings as delivered by the Prophet Muhammad (peace be upon him).

The main target of transformation in Islamic thought is to revitalise the good values and valuable tenets of Islam into practice. Resources in Islamic culture and literature provide the decent researcher with enough assistance in this endeavour. There is plenty of room for the reformation of legal, as well as social trends. Although, the latter might be much more difficult to uphold.

One of the major topics of the old *kalām* was the question of divine attributes. This was one of the main sources of disagreement and conflicts among the Mutakallimūn themselves, as well as with other theologians. For this reason, a new understanding of divinity was required. Allah is described in the Qur'an as: *He who nothing resembles him, yet He is the one who hears and sees.* This means that the reality of the Divine stands in His existence and His ability to act. His perception of things is a matter that goes beyond our comprehension, since he is not a material entity, and, therefore, we cannot expect Him to perceive through any physical means. Thus, this attribute must be taken *a priori* and not be subjected to rational analysis. Conversely, divine acts are something else, which we are directly concerned with. To understand the development of the world, our life and destiny in the presence of Divine action imposes the requirement that we should analyse and understand how God acts through his creations. The Qur'an calls upon us to contemplate creation and to question how it all started. Such an understanding will help us comprehend the mechanism by which the physical world developed and may help us understand our destiny. This can be seen as the role of analytic theology, whereby our understanding of creation may very much help us understand how our *re*-creation in the next life will be possible. If we adopt re-creation as a general rule, then it would be easy to envisage how the next form of life might be very different from this one. For example, the moment we cease to live, our souls may be re-created in another world. This assumption does not necessarily entail the many-worlds hypothesis, because the world into which the soul transfers is strictly non-physical.

To summarise: I would say that while the topic of divine action in this world is of interest to neo-*kalām*, the issues concerning divine attributes are not a matter of philosophical significance, but to be taken metaphorically without the need for further study. Nevertheless, spiritual experience should be allowed to have a room in neo-*kalām* as it constitutes a vital part of Islamic teaching and religious practice. The rational acknowledgement of spiritual experience can always be maintained because there is much to learn about our souls and the world which extends beyond our limited, current knowledge. That is to say, spiritual experience is something that cannot be denied. It is a real feeling, obtained through the interaction of our senses with their environment and our mind. Therefore, it cannot be but an integrated part of our consciousness which may well extend far beyond our direct sensing.

Chapter Two

The Kalām and the Mutakallimūn

In this chapter, I will describe the basis for *kalām* and its basic principles. In a brief historical review, I will present the sources of *kalām*, and its two parts: *Jalīl al-kalām* and *Daqīq al-kalām*, as well as the major schools of *Kalām*. I will end the chapter by listing the main principles of *Daqīq al-kalām*, which form the foundation for an Islamic approach to natural philosophy. But first, let me identify some basic definitions which will be used in this book.

WHAT IS KALĀM

In Arabic *kalām* means 'speech' or sequence of words; in the intellectual sense it is a 'dialogue' or a 'discourse.' In its philosophical context, *kalām* denotes a system of thoughts consisting of concepts, assumptions, principles, and problems which were used to explain the relationship between God, humans, and the physical world in accordance with the Islamic creed. Al-Sharīf al-Jurjānī (1339–1414) defines *'ilm* al-*kalām* as "the science that investigates the intrinsic properties of things in accordance with Islamic rules."[3]

3 Al-Sharīf al-Jurjānī, Kitab al-Ta'rīfat: *kalām*.

Classically, *kalām* was considered the foundation of jurisprudence (*fiqh*), which in turn constitutes the basis for Islamic Sharīʿa laws—the Islamic religious rules of life. The Sharīʿa laws represent a deductive system of rules and instructions which require a logical foundation, in addition to the sacred narrations, to be fully justified and established.

The discourse of *kalām* has been largely neglected in the last few centuries. This neglect was primarily due to Muslim scholars who found that *kalām* did not lead to fruitful results, especially in the context of classical natural philosophy. Furthermore, an attack, led by Ibn Taymīyyah, devalued the traditional contributions of the Mutakallimūn by refuting their rational arguments and considering *kalām* as corrupting to proper belief. This latter case was a strong and influential factor for many scholars in leaving the study of *kalām*.

However, having reviewed the propositions of *kalām* concerning natural philosophy, I feel that it has much to offer to the fields of natural philosophy and the present debates in science and religion, and is, therefore, worth studying. This book uncovers the scientific value of the doctrines of *Daqīq al-kalām* and may help to provide the basis for a philosophy of science to resolve many questions raised by modern physics. Many of the arguments of *Daqīq al-kalām* are still being vocalised and have a sound value in today's scientific studies, as well as in the philosophy of modern science. Indeed, the '*Kalām* Cosmological Argument', which was re-devised by William Craig,[4] is just one contemporary example in a field of ideas, concepts, and arguments that can be utilised by the modern philosophy of science. However, the subject is in such a state now that it cannot lend itself to an effective role unless it is first purified, reformulated, and harmonised to fit in with categories of modern philosophy. Much work and painstaking effort must be expected before *Daqīq al-kalām* is fit for a current role.

The terms *Daqīq al-Kalām* and *Jalīl al-kalām* have been mentioned by several pioneering scholars of *kalām*. For example, al-Ashʿari in his famous book, *Maqālāt al-Islāmyyīn wa Ikhtilāf al-Muṣallīn,* explicitly divides the topics of *kalām* into *Daqīq al-kalām* and *Jalīl al-kalām.*[5] Also, al-Naysabūrī and Ibn Matawayh, both Muʿtazilis, have mentioned this classification.[6] However, it might be illuminating to mention that the later Mutakallimūn did not use such classifications frequently.

Traditionally, the general trend on discussing *kalām* problems does not differentiate between questions of natural philosophy (physics) and those of

4 Craig, W. L., *The Kalām Cosmological Argument*, (London and Basingstoke: The Macmillan Press Ltd. 1979).

5 Maqālāt, p.300 and 301.

6 Al-Naysaburi, Masảil; Ibn Matawayh, Tathkirah.

philosophical or religious interest. It is also true that the late Mutakallimūn did not use this classification of *kalām* topics, despite being mentioned by al-Ashʿari and others. Nevertheless, I feel that in re-forming *kalām*, we need to revive such a classification, so that we can structure it as two parts: the Natural Philosophy part, and the Theology part. With this classification, we can re-structure *kalām* in a way that enables us to tackle problems more lucidly and base the new *Jalīl al-kalām* on a firm scientific basis.

For the sake of acquainting the reader with the necessary background in *kalām*, I am going to present and discuss those views that have a sound value in present-day natural philosophy. These will include my own re-arrangement and designations for the basic doctrines and principles of *kalām*. I will try to summarise their contributions to natural philosophy, which are covered under *Daqīq al-kalām*, and explain some vital problems where I feel genuine research needs to be done in order to identify the possible scope for deploying *kalām* in scientific and religious debates.

REASONS FOR THE RISE OF KALĀM

Historically, there are two basic motivations for the emergence of *Kalām*. The first was internal and concerned with rationalising the theological view within an Islamic framework. For example, the question of whether the sinner is to be considered a non-believer (*kāfir*) or not, triggered many discussions and arguments. Other questions were related to divine attributes, rewards and punishment in the afterlife, and man's freewill and responsibilities for his own deeds. These questions were part of *Jalil al-Kalām*.

The second reason for the emergence of *kalām* was the reaction of Muslims to the new ideas they encountered when coming into contact with new nations and civilisations, particularly the classical Mediterranean and Indic ones. This contact, at a time when Muslims were the dominant global power, created a 'dialogue between civilisations' rather than a 'clash of civilisations.' Muslims tried to justify their beliefs in a rational way, presenting sound arguments based on a worldview which took into consideration the construction of the world from basic elements and its physical properties.

THE TWO MAIN SCHOOLS OF KALĀM

The Mutakallimūn (doctors of *kalām*) formed two main schools: the *Muʿtazilis,* which was the first to be formed in Basra, and the *Ashʿaris,* which was formed about two centuries later by Abu al-Hasan al-Ashʿari (874–936), in Iraq. At the same time, a third school was founded in Samarqand and Khurāsān (present day Afghanistan). That school, called the *Maturīdī,* shared many of its views with the Ashʿaris; being mostly concerned with theological matters. The pioneering leaders of the *Mu'tazilis* were: Wāṣil Ibn 'Atā (d. 748), Amr Ibn 'Ubaed (d. 762), Abu al-Huthayl al-Allāf (d. 840), Ibrāhīm al-Naẓẓām (d. 835), and al-Jāḥiḍ (d. 868 A.D).

Much of the original works of the *kalām* leaders have been lost, but some of their key ideas and arguments have been passed down through the writings of their students and opponents. A number of valuable monographs and critiques have been preserved of leaders of the Muʿtazilis who lived during a later period. Most prominent among these were: Abu al-Hussein al-Khayyāt (d.-912), Abu al-Qasim al-Balkhi (sometimes called al-Kaʿbi) (d. 931), Abu Ali al-Jubbā'ī (d. 915), and his son, Abu Hāshim al-Jubbā'ī (d. 933). Some of the original works of these Muʿtazilis were maintained through monographs written by scholars such as 'Abd al-Jabbār al-Hamadānī (d. 1024), who wrote an extensive encyclopeadia about the Muʿtazilis. A student of his, Abū Rashīd al-Naysāburi (d. 1048), wrote about the different views of Baghdādī and Basrian Muʿtazilis, whilst another, al-Ḥasan Ibn Mattaweyh (d. 1059), relayed a good number of the opinions of early Muʿtazilis on *Daqīq al-kalām.*

The Ashʿari school was formed by Abū al-Ḥasan al-Ashʿari (d. 935), who broke away from the Muʿtazilis and formed a new school of thought within the wider circle of *kalām.* In addition to al-Ashʿari, the most prominent contributors to his school were Abū Bakr al-Bāqillānī (d. 1012), and Abu al-Maʿālī al-Jūwaynī (d. 1085), who wrote some excellent monographs on both *Daqīq al-kalām* and *Jalīl al-kalām.* However, one can say that the most efficient utilisation of *kalām* was made by al-Ghazālī (d. 1111), whose works represented the most mature writings produced by the Ashʿaris. In later times, the Ashʿaris *kalām* was reformulated by Saif al-Dīn al-Āmidī (d. 1233) and 'Aḍud al-Dīn al-Ījī (d. 1355); who was considered the last classical practitioner of *kalām.*

Daqīq al-kalām investigated some of the basic concepts of contemporary physics, such as space, time, matter, force, speed, heat, colour, smells, etc. So, it is quite legitimate to revisit this discipline, seeking a common understanding, not necessarily with physics as such, but perhaps with the philosophy associated with scientific concepts. This policy is supported by the fact that the resources of *kalām* are quite different from those of classical natural philosophy, including Greek philosophy.

The Mutakallimūn considered the Qur'an to be the primary source for their knowledge about the world and, consequently, they intended to establish their discipline in such a way as to understand the world according to its stipulations. This is the main reason we find some concepts of *kalām* differing in their meanings and implications from their apparent counterparts in both Greek and Indian philosophy. For example, the Qur'an stipulates that the world was created by God at some finite time in the past. Accordingly, the Mutakallimūn projected this condition into their theory of creation of the world and generated an understanding of the essences (*jawāhir*: singular *jawhar*) and transient attributes (*aʿrāḍ*: singular *ʿaraḍ*) as a general principle of discreteness (atomism) to serve this. On the other hand, for God to be free in designing the world according to His will, and in order that He exert full control over it, the world had to be thought of as being composed of a series of unstable and ever-changing events. This requirement produced the concept of ever-changing properties (called transients), which was expressed in the principle of continued re-creation. This led theologians to consider the results of the laws of nature (fire burning cotton, for instance) as being undetermined, such that the Mutakallimūn were able to develop a new concept of causality.

In no way do I wish to claim here that *kalām* formed an integrated body of thought, or that it can be associated with any individual Mutakallim (practitioner of *kalām*), or that it gives a complete, and satisfactory, explanation for modern philosophy of nature. Rather, I will try to uncover those ideas of the Mutakallimūn which might serve as candidates for integration within contemporary philosophies of natural science. For example, the principle of continual re-creation can be utilised to better understand the state of indeterminacy of measurement in the physical world. Also, the notion of 'discrete time', which was proposed by the Mutakallimūn as part of the general principle of discreteness (atomism) in nature, can be applied to construct an 'all-discrete' theory of nature that might help eliminate the current fundamental theoretical problems associated with the unification of natural physical forces.

On the other hand, some questions that had been considered by *Jalīl al-kalām* resonate with contemporary debates in science and religion which are taking place in the West. Questions concerning the knowledge of God, His action in the physical world, His control of the future, and the degree of freedom enjoyed by the natural world and humans, were some of the issues debated by the Mutakallimūn.

SOURCES AND METHODOLOGY OF KALĀM

Ibn Khaldūn, the famous historian of thought and sociology, tells us in his book, *al-Muqaddimah,* that the early Muslims in trying to explain articles of faith, initially quoted verses from the Qur'an and provided narrations of the Prophet. Later, when differences of opinions occurred concerning the details of these articles, "argumentation formed by the intellect ('aql) began to be used in addition to the evidence derived from tradition, and in this way the science of *kalām* originated."[7]

7 Ibn Khaldun, Al-Muqaddimah, p.

The Mutakallimūn considered the Qur'an to be their main source for deducing knowledge about the world. Although they did not explicitly refer to Qur'anic verses, it was clear that their main principles were deduced from it. Hence, they followed a logical sequence of deduction which started with divine revelation, and then tried to explain nature accordingly. Richard Walzer summarised this by saying that the "Mutakallimūn follow a methodology that is distinct from that of the philosophers in that they take the truth of Islam as their starting point."[8] William Craig has taken the same view, saying that "the main difference between a mutakallim [practitioner of *kalām*] and a *failasūf* ['philosopher'] lies in the methodological approach to the object of their study: while the practitioner of *kalām* takes the truth of Islam as his starting-point, the man of philosophy, though he may take pleasure in the rediscovery of Qur'anic doctrines, does not make them his starting-point, but follows a method of research independent of dogma, without, however, rejecting the dogma or ignoring it in its sources."[9]

The approach of the Mutakallimūn in understanding the world can be presented as follows:

The World $\Rightarrow$ Reason $\Rightarrow$ God

This is the opposite to the approach of Greek philosophers, which can be presented by the sequence:

God $\Rightarrow$ Reason $\Rightarrow$ The World

Effectively, this same difference applies to the Mutakallimūn and Muslim philosophers, who tried hard to reconcile Greek philosophy with Islam.

PHILOSOPHY AND KALĀM

As mentioned above, most Muslim philosophers tried to accommodate Greek philosophical trends within the Islamic worldview. This reconciliatory approach was started by al-Kindī (d. 868 A.D), and further developed by al-Fārābī (d. 950) and Avicenna (Ibn Sīnā, d. 1036), who adopted a mainly neoplatonic approach. The practise of the early Muslim philosophers in recognising divine action in the world was refuted by al-Ghazalī in his classic, *The Incoherence of the Philosophers*[10] (*Tahāfut al-Falāsifah*). In turn, Averroes (Ibn Rushd, d.1198) later countered al-Ghazālī's arguments in his *Incoherence of the*

8 Richard Walzer, 'Early Islamic Philosophers', in A. H. Armstrong (ed.), *The Cambridge History of Late Greek and Early Medieval Philosophy*, (Cambridge: Cambridge University Press, 1970), p. 648.

9 Craig, p. 17 and references therein.

10 Abū Ḥāmid al-Ghazālī, *Tahāfut al-Falāsifah*, (*The Incoherence of the Philosophers*), translated by Michael Marmura (Provo, Utah: Brigham Young University Press, 2000).

Incoherence[11] (*Tahāfut al-Tahāfut*) and defended Aristotle's doctrines. Moreover, in *Faṣl al-Maqāl* (Decisive Treatise), Averroes strove valiantly to show that Islam can accommodate Greek philosophy through certain reinterpretations of the verses of the Qur'an.[12] However, this defence was unsuccessful, since the arguments presented by al-Ghazālī were strong and proved very effective in persuading the elite of the inadequacies of philosophising.

The fact that, at the time of al-Ghazālī, *kalām* was still under siege and frequently out of favour with many religious scholars and jurists, caused one branch of Islamic thought (the Ḥanbalī) to be directed towards a more fundamentalist approach which bred hard-line thinkers, such as Ibn Taymīyyah (d.1328). The birth of trends that minimised the role of a rational approach in understanding God and the world did not assist the growth of reason-based theology or science in the Islamic world.

Some of the Mutakallimūn, who lived during the 11th century and later, especially those whose allegiance was still to the Muʿtazilis, borrowed some of the philosophical arguments in their endeavours to support the existence of God and theorise His attributes coherently. This approach was not an unmitigated success, for it appeared inconsistent with the basic *kalām* thesis, which assumed that revelation was the prime source of knowledge and enabled us to understand the world.

THE TRADITIONAL METHODOLOGY OF KALĀM

The Mutakallimūn relied much on the logical approach they adopted. This logic did not follow Aristotelian formal logic. Instead, it was a simpler methodology that approached a problem according to certain logical categories, examined the possible options, and excluded any that did not hold up to their rational analysis, until they reached a conclusion. This method was called 'analysis and classification', (al-sabr wa al-taqsīm). As such, they had their religious belief driving their evidence and choices of solutions. This is similar, in essence, to what we nowadays call the method of 'conjecture and verification'; a very useful technique when dealing with questions of natural philosophy. Ibn Rushd, following Aristotle, classified philosophical arguments into three types:

Rhetoric: which uses the power of linguistic construct to convince the reader or listener without necessarily containing any logical reasoning or proof. Such 'evidences' are routinely used by religious preachers.

---

11 Averroes, Tahāfut al-Tahāfut, (*The Incoherence of the Incoherence*), translated from Arabic with introduction and notes by Simon Van den Bergh (London: E.J.W. Gibb Memorial, 1930).
12 Ibn Rushd, *Faṣl al-Maqāl*, published and translated as: *The Philosophy and Theology of Averroes*, trans. Mohammed Jamil-al-Rahmān (Baroda: A. G. Widgery, 1921).

Dialectic: These were mostly used by the Mutakallimūn, and demonstrate a clear, rational approach towards proving an argument. It provides a well-defined premise and a logical construct that follows up its reasoning in several steps. Al-Ghazālī provided some very rich material regarding these approaches in his books, *Miʿyar al-ʿIlm fi al-Mantiq* (The Standard of Science in Logic), and *Maḥak al-Naẓar* (The Challenge of Thought). Among those Mutakallimūn in whom we find a wide use of the systematic, dialectic approach, is Abū Bakr al-Bāqillānī of the Ashʿaris and Ibn Matawayh of the Muʿtazilis.

Demonstrative: This type of proof is considered the highest ranking when it comes to presenting an argument and has the strongest methodology for augmenting a case. Such evidence widely uses arguments of formal logic, as well as geometrical and mathematical proofs. Both al-Bāqillānī and al-Ghazālī used demonstration when presenting their views, and we find the latter providing some very sophisticated geometrical evidence in his book, *The Incoherence of the Philosophers*. In addition, Ibn Ḥazm (994-1064) presented geometrical evidence and examples from astronomy to support his arguments on the temporality of the world.

Ibn Rushd (1126–1198) praised using the demonstrative approach in philosophy, even though some of his demonstrative arguments were not strong enough to stand up to the scrutiny of the Mutakallimūn. We see an example of this in discussions about the size of the world by al-Ghazālī, and whether it could be larger or smaller than it actually was. Here, we find Ibn Rushd in difficulty over presenting any convincing evidence and resorting to the Aristotelian view, which says that any change in the size of the world would spoil it. This is a weak and unsubstantiated claim. Another example is the argument presented by Ibn Rushd in defence of Galen, concerning the degeneration of the sun and his denial that it could corrupt. These examples are explained in detail in Chapter Eleven of this book.

THE ROLE OF ARABIC IN ISLAMIC KALĀM

From advanced linguistic studies, we know that language is not only a communication means but also a cognitive tool. The view that language is simply a complex communication system that does not influence cognitive processes in any substantial way has been criticised from several perspectives. A new framework, called Embodied Cognition, considers cognitive processes as "non-symbolic and heavily dependent on the dynamical interactions between the cognitive system and its environment."[13] Studies on this topic date back to the 1930s, with the work of Russian scholar, Lev Vygotsky.[14]

13 Mirolli, M. and Parisi, D., "Language as a Cognitive Tool, Minds and Machines", Volume 19 Issue (4) 2009 pp 517–528.
14 Vygotsky, L.S. (1978), *Mind in society*, Harvard University Press, Cambridge; (1962),

Recently, the idea of language as a cognitive tool has been given attention within the cognitive-science-oriented philosophy of mind (see, for example, Carruthers and Boucher).[15] Daniel Dennett[16] has argued that the human mind, including consciousness, depends mostly on the way human brains are substantially 're-programmed' by cultural input which is derived, principally, through language. Andy Clark[17] has further developed Dennett's ideas to conclude that language is not only a communication system, but also a kind of "external artifact whose current adaptive value is partially constituted by its role in re-shaping the kinds of computational space that our biological brains must negotiate to solve certain types of problems, or to carry out certain complex problems." Surely, the influence of language on the cognitive system increases, as long as language is rich and versatile enough to accept descriptions and various aspects of things. The richer the language is with regards to synonyms, the more it can contribute to developing an advanced cognitive system.

Arabic is a strong and very rich language. It is well-structured and has the property of lending its terms to derivation. This important property can contribute to building a deductive system of thought. Therefore, it is not surprising to learn that Arabic has played an important role in formulating Islamic thought. This has been implicitly acknowledged by the Qur'an where it speaks of the role of Arabic in casting its content, "Thus have We revealed it to be a judgment of authority in Arabic," (13:37) and we also read, "Surely, We have made it an Arabic Qur'an that you may understand." (43:03). In the second verse, the phrase: *we have made it an Arabic Qur'an,* implies that it contains the spirit, or way of thinking, of the original Arabic. This understanding becomes clearer when we read the verse, "And if We had made it a foreign Qur'an, they would have said: Why were its messages not elaborate? What! a foreign and an Arab! Say: It is to those who believe a guidance and a healing, and those who believe not, there is a deafness in their ears, and it is obscure to them." (41:44).

In formulating the concepts of *Kalām*, I can say that Arabic played an essential role, much like the role played by mathematics in physics. When a

---

Thought and language, MIT Press, Cambridge, MA.

15 Carruthers, P. & Boucher, J., ed. (1998), Language and thought: Interdisciplinary themes, Cambridge University Press, Cambridge.

16 Dennett, D.C. (1995), Darwin's Dangerous Idea: Evolution and the Meanings of Life, Simon and Schuster, New York, NY; (1993), 'Learning and labelling', Mind and Language 8(4), 540–547. Dennett, D.C. (1991), Consciousness Explained, Little Brown & Co., New York, NY.

17 Clark, A. (1997), Being There: putting brain, body and world together again, Oxford University Press, Oxford; (1998), Magic words: How language augments human computation, in Peter Carruthers & Jill Boucher, ed., 'Language and thought: Interdisciplinary themes', Cambridge University Press, Cambridge, pp. 162–183; (2006), 'Language, embodiment, and the cognitive niche', Trends in Cognitive Sciences 10(8), 370–374

term is spelled out in Arabic, it generates a new conceptual derivative through its linguistic derivatives. The role of Arabic in creating concepts of *kalām* can be easily recognised through the expressions of these terms and their Arabic meanings in relation to their conventional meanings in the language of *kalām*. For example, the word *'araḍ* is the true word for something transient and for something that does not endure. Take, for example, the Qur'anic description of the cloud, "this is *'āriḍ* which will bring us rain" (43:24). The word *'āriḍ* (adjective for *'araḍ*) describes the cloud as a *transient* and non-enduring entity. The same term, with the same meaning, is used in *kalām* to describe the non-enduring properties associated with the essence (the *jawhar*).

On studying *kalām* one should differentiate between two types of cognitive origins or influence: linguistic and dogmatic. The linguistic influence comes through language itself in relation to a cognitive system built over a long period of time, while the dogmatic origin comes from the system of belief represented by Islam.[18] The fact that they are extensively interwoven does not allow much room for differentiation, but certainly one can recognise such roots when a piece of text is subject to linguistic analysis and scrutiny. Truly, both origins interact and influence one another in such a way as to create the spirit of the nation. In this respect, both Islam and Arabic have a dialectic relationship (in the Hegelian sense) that has provided us with a wealth of cultural and intellectual material, which has its imprint in world culture. As for the logic build up, we should also know that this is not independent of culture and dogma but is deeply rooted in the cognitive system for which humans have been picked amongst nature.

The biggest problem facing *kalām* was the problem of interpretation. Understanding the meaning of the verses of the Qur'an caused disagreements between different sects of the Mutakallimūn. In fact, this very reason was behind the appearance of divergent and conflicting schools of thought in Islam. The main issue upon which those schools differed were the attributes of the Divinity and human freewill. The question of God's attributes uncovered serious differences and led to much debate amongst the Mutakallimūn. The Mu'tazilis adopted a rational approach and held a realistic view from which they addressed the question of divine attributes. They found the Qur'an identifying God (Allah) as an entity not resembling any other in existence. Rationally, this is obvious, as Allah is non-physical. Accordingly, they interpreted the verses which assign physical attributes, such as having limbs, hands, eyes *etc.*, as being metaphors, conveying certain supreme qualities of Allah that He expressed

18 This does not necessarily mean to say that Islam is a cultural product of the Arabs, despite the strong ties between both. Islam, as delivered through the Qur'an, is composed in the Divine word, as expressed in Arabic.

in the Qur'an as resembling our qualities. The same applies to the abilities of Allah, such as hearing and seeing. The Mu'tazilis asserted the uniqueness of Allah, his omnipotence and omniscience, without comparing it to our human capabilities and the means by which we learn, hear and see. Thus, Allah is an independent entity with nothing to resemble him. By the same token, when some of the Mu'tazilis thought about the question of seeing Allah, they denied that the Divine could be seen or communicated with, even in the *Ākhirah* (the afterlife). But in this case, they put themselves in an embarrassing situation, as the Qur'an speaks openly about the ability to see God in the *Ākhirah*. However, I do not want to get into this discussion, other than to say that this analogy between man and God is motivated by the image of a personal God, the physical *dunyā* (this life) and the metaphysical *Ākhirah,* which are dogmatic problems that we find in *Jalīl al-kalām*. A much more accurate and proper understanding of divinity, and questions related to it, comes once we acknowledge that Allah is neither a physical body, nor a part of our material world. He is not a corporeal object that can be touched and dealt with by physical means. Consequently, questions, such as whether we are able to see Allah in this Dunyā, are substantially different from whether we can see him in the *Ākhirah*, since our state in the afterlife is fundamentally different from our state in this life. Those contemporary theologians of Islam who do not realise these facts will not be able to produce any fruitful lines of thought that are compatible with the current state of knowledge and intellect.

OUR AUTHENTIC SOURCES

Much of the intellectual work on *kalām* was lost, either unintentionally or intentionally, during the reign of the Abbasid Caliphs, al-Mutawakkil (847–861) and al-Qādir (991–1031). These two Caliphs banned the Mu'tazilis from teaching their views and practising their way of studying. They were replaced by their Ḥanbalī rivals, who addressed theological questions using a less rational and more literal narrative approach.

Books of the *kalām*, especially those written by the Mu'tazilis, were burned, with only a few escaping destruction. Among these are the books of Qaḍī 'Abdul Jabbār, who wrote a multivolume encyclopaedia of Mu'tazili thought and scholars, entitled *al-Mughnī fi abwab al-Adil wa al-tawḥīd*. He is also thought to have penned *Al-Muḥīt bil Taklīf,* although this book is claimed to have been written by his famous student, al-Ḥassan Ibn Matawayh. This man was an exponent of the Mu'tazilis and wrote an important book on the subject of atomism in *kalām*. Another well-known student of 'Abdul Jabbār's was Abū Rashīd al-Naysabūrī, who wrote about the problem of disagreements between the Basriyyan and Baghdādī Mu'tazilis in *al-Mas'l fi al-Khilaf bayna al Basriyyīn wa al-Baghdadiyyīn* and *Dewan al-Uṣūl*. Also, from the saved legacy of the Muʿtazilis, we have the book of al-Khayyāt, in which he refutes the claims of Ibn al-Rawindī, an opponent of the Muʿtazilis. This book is significant because it contains many quotations attributed to Ibrahim al-Naẓẓām, of whom we do not have any saved work. In the well preserved literature of al-Jāḥiẓ, we also find some of the ideas and views of al-Naẓẓām, who was his mentor. Most of the views of the Muʿtazilis have been preserved through the writings of the Ashʿaris, primarily in the famous book of Abū al-Ḥasan al-Ashʿari entitled, *Maqālāt al-Islāmiyyīn wa ikhtilāf al-Muṣallīn*. This is, perhaps, the most reliable book on the legacy of *kalām*, and is the main source of our knowledge about the views of the early Mutakallimūn. In it, al-Ashʿari separates the topic of *Jalīl al-kalām* from *Daqīq al-kalām*, while most other books mix the two.

Records written by the Ashʿaris are more readily available. For example, we have the books of Abū Bakr al-Bāqillānī, the famous Mutakallim. The most specialised of these is *Kitab Tamhīd al awa'il wa Talkhīs al-Dala'il*, in which he presents the views of the Ashʿaris in respect of *Daqīq al-kalām*. This theologian is one of the pioneering Ashʿaris who contributed substantially to theorising the arguments of this school of thought. Additionally, we have the works of Abū al-Maʿālī al-Jūwaynī, an important one of which is entitled, *al-Shāmil fi Usūl al-Dīn*. This volume deals mostly with the doctrines of *Daqīq al-kalām* and is a prime source for our knowledge on *kalām* atomism. Al-Juwaynī's *al-Irshād*, on the other hand, is mostly concerned with *Jalīl al-kalām*.

Other works by theologians who are not considered to belong to the Mutakallimūn, but in some way subscribe to *kalām* or share some views of it, are also available. Among these is that of Ibn Ḥazm, the theologian of Granada, entitled *al-Fiṣal fi al-Milal wa al-ahwā' wa al-Niḥal*, in which he presents criticisms of the different factions and schools of thought, among which are the Mutakallimūn. Although Ibn Ḥazm is not counted among the Mutakallimūn, some of his ideas and opinions reflect the influence of *kalām* and are profoundly connected with it, especially those concerning the principles of *Daqīq al-kalām*. As such, Ibn Ḥazm has contributed so much to defending the Ashʿaris 'views, that, aside from the historical classification of thought, I would certainly consider him as subscribing to the Ashʿaris school.

A similar case is found with Abū Ḥāmid al-Ghazālī, who is not officially counted as a Mutakallim, but who has contributed a great deal in defence of many of the *kalām* doctrines, especially those related to topics on natural philosophy. His book, *The Incoherence of the Philosophers* (*Tahāfut al-falāsifah*), is a great work in which, in my opinion, he gives the best presentation of Ashʿari *kalām* with respect to topics on *Daqīq al-kalām*, such as temporality and natural causality. In contrast, we find that the book of Ibn Rushd, *The Incoherence of the Incoherence,* provides an excellent source for the Aristotelian response to the arguments of al-Ghazālī. Ibn Rushd has other important books, such as *al-Kashf 'an Manāhij al-Adilla,* in which he discusses and criticises the basic doctrines of *kalām*, including atomism.

Books written by several biographers and historians of thought, such as the renowned Sunni jurisprudent, 'Abd al-Qahir al-Baghdādi (*al-farq bayn al-Firaq*) and the historian of *kalām*, 'Abdul Karīm al-Shahrustānī (*al-Millal wal-Nihal* and *Nihāyat al-Iqdām fi 'ilm al-kalām*) are good sources of historical matters concerning the different sects of Muslims who practised contemplation. However, they should be read with caution as they might be less reliable, especially al-Baghdādī's works, which contains biased attacks on other factions.

One more clarification is due, which is that the studies described in this book do not use sources of *kalām* from the later Mutakallimūn, i.e. those who came after Abū Hamid al-Ghazālī. Instead, they are mainly concerned with sources from the early Mutakallimūn, who contributed their thoughts during the first five centuries of Islam. The reason is that I find more originality with the early Mutakallimūn; feeling that they were more attached with the spirit of the Qur'anic methodology of investigation. The later Mutakallimūn, starting with al-Fakhr al-Rāzī (1149–1209), leaned towards the philosophical approach, and perhaps al-Rāzī more so than others since, according to certain measures, he confused philosophical views with *kalām* to the point of negating the basic principles of *kalām* in respect of *Daqīq al-kalām*. For example, his position is not clear about *kalām* atomism, and I am not sure of his stance on the question of causality and causal determinism.

In order to double-check my attitude and judgment about the later Mutakallimūn who appeared after al-Rāzī, such as al-Āmidī (1156–1233), al-Ijī (1281–1355) and al-Taftazānī (1322–1390), I examined a large portion of their works and found them less influenced by philosophical arguments than al-Rāzī.[19] However, I also say that nothing original was found in them that was worth considering for this book. Nevertheless, wherever I found it essential, their useful contributions have been noted. For this purpose, I tried to set assessment measures when considering a philosopher or Mutakallim; the target being to apply Ibn Khaldūn's statement about these later Mutakallimūn, in which he says, "The later scholars were very intent upon meddling with philosophical works. The subjects of the two disciplines (theology and philosophy) were thus confused by them. They thought that there was one and the same (subject) in both disciplines, because the question in both disciplines were similar."[20]

19 M.B. Altaie, *The Impact of Philosophical Approach on Late Kalām*, Istanbul, Dec.2015
20 Ibn Khaldūn, *al-Muqaddimah*, translated by Franz Rosenthal in three volumes, (Boston: Princeton University Press 1967) vol. III p. 35–40.

I have benefited from studying several modern analyses on *kalām* conducted by Western Orientalists. One of these was Donald Black MacDonald, whose views I found harsh when it came to evaluating some of the ideas and concepts of *Daqīq al-kalām,* as he ignored important issues concerning the motivations behind early *kalām* discussions. I have also learned a good deal about basic concepts and different uses of terms related to atomism from Slomo Pines' excellent book, *Studies in Islamic Atomism.* In it, Pines analysed the origins of *kalām* atomism and concluded that it cannot be attributed to a Greek or Indic source. Also, I found the study of Pretzil[21] useful, despite not agreeing with his conclusions, which had previously been refuted by Pines. *The Philosophy of the Kalām* by Harry Wolfson stands as a monumental work on *kalām*. The book discusses these concepts in a historical context and searches for a Greek or Indic origin. Al-Noor Dhanani's Ph.D. thesis entitled, *The Physical Theory of Kalām*, is a presentation of topics in *Daqīq al-kalām,* mainly related to their atomic theory. The presentation is well-arranged and contains a large number of excerpts from the original books of *kalām*, such as al-Ashʿari's *Maqālāt*, al-Naysābūrī's *Masāʾil,* and Ibn Matawayh's *al-Tadhkirah,* to substantiate his analysis. However, Dhanani mis-expressed some of the original *kalām* terms when translating them into English, causing some serious distortions in the text, which may affect their actual meaning. My critical assessment of some of these expressions, and an understanding of these terms, will follow in the course of this book.

As far as modern Arabic studies of *kalām* are concerned, one notices the poor content on the Arabic shelves in libraries on this topic. Only a few serious studies can be found. These have been mostly conducted by students of Western Orientalism and are mainly concerned with *Jalīl al-kalām*. Ḥusam al-Alūsī was one of the serious scholars who contributed quality work on the problem of creation. This was part of his Ph.D. dissertation submitted to Cambridge University during the mid-1960s. His work came under the title, *The Problem of Creation in Qur'an, Hadith and Commentaries*. Following on from his doctorate, Ḥusam al-Alūsī published many excellent books detailing the conflict between the philosophers and the Mutakallimūn on certain issues concerning causality and creation. Additionally, Yumnā al-Khūlī has published a good treatise about the physical theories of *kalām* in a book entitled, *Tabīʿiyyāt al-Kalām*, which I found very interesting and useful for those beginning their studies on *Daqīq al-kalām*. Furthermore, Muna Ahmad Abu Zaid has published an excellent piece of historical analysis on *kalām* atomism in a book entitled, *The Atomic Visualization in Islamic Thought* (*al-Tasawūr al-Tharrī fi al-Fikr al-Islamī*), in which she quotes widely from original sources of *kalām*.

21 Perzil, O., Die fruhislamische Atomleher, Der Islam,19, (1931), 117–30.

THE MAIN PRINCIPLES OF *DAQĪQ* AL-KALĀM

At the outset of this topic, we should recognise the fact that the worldview constructed by the Mutakallimūn was the result of their belief, and their rational analysis of those beliefs. The Mutakallimūn's understanding of the world and their interpretations of various events basically stemmed from their religious belief. God is one, unique and not a multiple entity. God is omniscient, omnipotent, and just. Based on the stipulation of the Qur'an about the world and man, the Mutakallimūn construed their doctrines. Out of those basic principles, expositions, or premises, the Mutakallimūn developed their view of the natural world. Even though the Mutakallimūn expounded diverse views according to their school affiliation, one finds that almost all of them subscribed to certain common principles when it came to understanding nature. I have studied the *kalām* legacy, specifically that part dealing with questions on natural philosophy, and was able to identify five basic principles.[22] Here, I will give a brief description of the principles which will be presented in much more detail in Part II, which is devoted to presenting and discussing their implications. These are:

1. **Temporality**. According to the Mutakallimūn, the world is not eternal but was created at some finite point in the past. Space and time had neither a meaning nor existence before the creation of the world. Despite the fact that some of the Mutakallimūn believed that creation took place out of a pre-existing form of matter, the dominant view of *kalām*, in this respect, was that creation took place *ex nihilo*, that is to say, out of nothing. Accordingly, they considered every constituent of the world to be temporal.
2. **Atomism**. The Mutakallimūn believed that all the entities in the world are composed of a finite number of a fundamental element, called *jawhar* (essence), which is indivisible and has no parts to it. The *jawhar* on its own is thought to be an abstract entity that acquires its physical existence and value when it becomes associated with a character called *'araḍ* (a transient attribute). At least one *'araḍ* is necessary, and these *a'rāḍ* (plural for *'araḍ*) specifying the *jawhar* possess ever-changing qualities. A *jawhar* and at least one essence can form the element of composition of things. This is known as the Islamic atom, or an elementary particle of the composition of the world. Discreteness applies not only to material bodies, but also to space, time, motion, energy (heat), and all other entities. The most important property of the *jawhar* is the occupancy and acceptance of the transient attributes.

22 M.B. Altaie, 'The scientific value of *Daqīq* al-*Kalām*', Islamic Thought and Scientific Creativity, Vol. 4, No. 2 (1994), pp.7-18.

3. **Continual re-creation.** Because God is believed to be the absolute able Creator of the world, and because He is living and ever-acting to sustain the universe, the world has to be continuously re-created. This mechanism is thought to be necessary for the existence of the universe and the validity of Divine action in nature. This continued re-creation occurs with the transient attributes of elements and not their essence. However, since the essence cannot be realised without being attached to a transient, the re-creation of the transient effectively governs the essence too. By such a process, God stands to be the sole Sustainer of the world. This principle is very important for two reasons: firstly, it establishes an indeterminate world, and secondly, it finds a sound, physical echo in contemporary quantum physics, as will be shown later. This principle can provide a clear explanation to many of the problems in modern physics, which come together under the title, "The Problem of Measurement in Quantum Mechanics."[23]
4. **Contingency and indeterminacy of the world.** Since God possesses absolute free will, and since He is the personal Creator and Sustainer of the world, He is at liberty to take any action He wishes, with respect to the state of the world or its control. Consequently, the laws of nature that we observe, and not the laws of physics that we devise, must be probabilistic rather than deterministic, so that physical values remain contingent and undetermined. From this theological background, the Mutakallimūn deduced the indeterminacy of the world. This resulted in their rejecting the existence of natural causality because nature, according to *kalām*, cannot possess any sort of will. The Mutakallimūn also rejected the Greeks' four basic elements and the existence of any kind of self-acting property inherent in those elements. This became a central argument in *kalām* as proof of a need for God; if nature is blind, then no productive development would be expected.
5. **Integrity of space and time.** The Mutakallimūn believed that space has no meaning on its own. Without there being a body, we cannot conceive the existence of space. This is also the case with time, which cannot be realised without the existence of motion, which in turn needs a body to be affected. For the Mutakallimūn, space and time are emergent properties that are realised upon the existence of matter and motion. This connection between space and time is deeply rooted in the Arabic language itself. Therefore, neither absolute space nor absolute time, in fact, exist. This belief drove their understanding of motion as being discrete, so that the trajectory of motion is composed of neighbouring 'rest-points' which are

23 M.B. Altaie, M. B. Altaie, Re-Creation a Possible interpretation of Quantum Indeterminism, in Matter and Meaning, Edited by Michael Fuller, (UK: Cambridge Scholar Publishing, 2010) p.21–36.

caused by the re-creation of the moving particle. Accordingly, they said that a body is seen moving faster than another only because the number of rest-points along its trajectory is small compared to those along the trajectory of the other. More boldly, the Muʿtazili, al-Naẓẓām, believed that motion on the microscopic level takes place in discrete jumps, called *ṭafra*. Max Jammer considered this understanding of al-Naẓẓām's to be the oldest realisation of the principle of quantum motion.[24]

THE STATUS OF THE PHILOSOPHY OF SCIENCE

In this section, I will try to give a short overview of the status of contemporary physics and its related philosophical questions. This may shed some light on the reasons why we need to have a new approach for Natural Philosophy. This introduction may help us understand the practical and pragmatic motivations for presenting the views of *Daqīq al-kalām* in the new context of modern physics.

The discoveries of the Theory of Relativity and Quantum Theory at the dawn of the last century subjected contemporary philosophy to a kind of shock and awe moment. Consequently, some radically new concepts emerged with which the philosophies of the 18th and 19th centuries were entirely unfamiliar. The most prominent of these new concepts was the integrity of spacetime, proposed by the Theory of Relativity, and the concept of indeterminism, which was required by quantum physics. The latter concept meant that it would be impossible to predict any natural development with complete certainty; a concept that jeopardised the Laplacian doctrine of determinism of the natural world, which had become one of the main pillars of classical natural philosophy of the 18th and 19th century. In fact, so strong was the shock, that philosophers had to pause for some time before accepting the new paradigm. Some philosophers continued to deal with the topic of natural philosophy using the same paradigm as the classical philosophy of science. However, any inspection of the *status quo* of contemporary science makes it clear that no viable philosophical framework for modern science can be acknowledged other than empiricism. The reason for this, is that the European classical Enlightenment philosophies could not deal with the conceptual development of modern physics in the 20th century.

Western philosophical heritage seems to be incompatible with the philosophical basis for modern physics, the Theory of Relativity and quantum physics. This is because Western classical philosophies were based on the fundamental doctrines and dialectical methodology of the Greeks, which were

24 Max Jammer, *The Philosophy of Quantum Mechanics* (New York: Wiley, 1974), p. 259.

comprised of a philosophical system, and had, at their core, fundamental principles which inclined towards a solidified view of nature without God. For this reason, I conjecture that the modern Western philosophies of science, with their roots still in Athens, have been unable to structure a consistent framework. Instead, modern philosophers of the 20th century, for example Karl Popper, simply denied the need for a framework. One might conclude that, in general, the Western mind has never acknowledged, philosophically, the development of abstract concepts in quantum physics. Most philosophers of science, including Popper himself, specifically critiqued the Copenhagen School's interpretation of quantum measurements. It might be fair to say that the Western mind has never accepted the indeterminism of natural phenomena and events. Indeed, theories were developed in order to avoid the indeterminism of quantum mechanics, claiming that a hidden determinism is the underlying truth in an apparently indeterministic world.[25] These so-called Hidden Variables Theories still struggle to defy the natural reality of indeterminism. However, the achievements of quantum physics are well established now, and Quantum Theory has proved to be a consistent theory, despite the possibility that it might need complementary postulates to explain some of its mechanisms. Such postulates, which I see in the re-creation hypothesis, and which once adopted many paradoxes, such as the double slit experiment, Schrödinger's cat, and the EPR 'spooky action at a distance,' can be resolved in quantum mechanics. Recently, a more appropriate concept of 'natural realism' was developed in order to reform the philosophical propositions of empiricism and to shape a more consistent and viable philosophy of modern science.

We need to understand the true implications of 20th century science as much as we need to understand its philosophical implications. Concepts, proposed by Quantum Theory and the mathematical structure of quantum mechanics, are still in need of a deeper understanding and interpretation. The meaning of an 'operator' in quantum mechanics is obscure, as is the meaning of the unpredictability of measurements. The role played by mathematical entities, called 'imaginary quantities' in physics, although being directly un-measurable, is worth studying on the same level as philosophical concepts, so as to understand the key dimensions of its practical, naturalistic meaning.

In theoretical physics, most of us physicists play the game of generating equations that do not always have a clear explanation. An example of this is 'string theory.' This theory has many versions which, according to some scientists, might be equivalent. However, the unproven status of this theory

25 This can be seen in the works of David Bohm and collaborators, see for example: Bohm, D., & Hiley, B. J. (2006). *The Undivided Universe: An ontological interpretation of quantum theory*. Routledge.

shows that no serious testable prediction has been found which would make this mathematical construct a viable theory of physics. In general relativity and curved space-time physics, we are not ready yet to understand the full meaning and implications of some concepts and phenomena, despite the solid, physical background of the Theory of General Relativity and the great success of its applications in cosmology. Nevertheless, some concepts, such as the space-like regions of space-time, remain obscure. Much of the theoretical problems in cosmology are due to our inability to find a common ground between quantum physics, which adopts indeterminism of events and discreteness of physical quantities, and the Theory of General Relativity, which adopts determinism and continuity. Some current problems of theoretical physics need the resolution of such a conflict in order to solve them. To understand the role of gravity at very short distances, we need to have a quantum theory of gravity; a long-sought after target that is not yet in hand. This includes understanding interactions in the microscopic world, which lies beyond the present limits.

Despite the great fame that Stephen Hawking gained during the past few decades, the physics of black holes is far from being understood. Part of the problem, again, is due to the incompatibility of quantum mechanics and general relativity. For this reason, many black hole physicists were taken by surprise by the declaration of Stephen Hawking's that information is not completely lost when a particle falls into a black hole.[26]

In cosmology, and despite the eminence of the Big Bang Theory, we are still far from deciding whether the universe had a starting point in time or whether it has an infinite extension into the past. The point of singularity, which contained all the matter and energy existing in our universe, stands, not only as an epistemological challenge, but as an ontological dilemma too. The problems of identifying 'dark matter' and 'dark energy' are a stumbling block to obtaining a clear understanding of the constituents of the universe. In fact, we are not yet sure whether our universe is one universe or many, or whether it is finite or infinite!

Some of the problems in theoretical physics, including that of measurement in quantum mechanics, stems from philosophical attitudes. Albert Einstein, believing in a deterministic world, shouted that, 'God does not play dice!' and rejected quantum indeterminism. But he was wise enough to devise an experiment from which he conjectured that, according to quantum mechanics, quantum states should be entangled. Since no such thing was known to exist back then, to Einstein this indicated that quantum mechanics was incomplete. In 1982, this entanglement was experimentally demonstrated and Einstein's claim about the incompleteness of quantum mechanics was shown

26 Stephen Hawking, GR17 conference, Ireland, public lecture.

to be flawed. There yet remains the question of giving a proper convincing interpretation for all this. A similar situation occurred when Ibn Rushd (Averroes) rejected atomism, claiming that adopting such a principle would turn geometry into numerology, and since this could not be done, he found it hard to accept atomism (see Chapter Three). Time proved Averroes correct in his conjecture, but wrong in rejecting atomism.

During the second half of the last century, a number of physicists, including Robert Dickie, Paul Davies, Martin Rees, turned their attention to the accuracy with which the values of universal constants (e.g., the mass of the electron, Planck's constant, the gravitational constant, the strength and range of nuclear forces, and the charge of the electron) have been set. They discussed how sensitive the structure and properties of our universe might be to changes in those values. The conclusion they came to was that a life-accommodating universe would be extremely sensitive to even small variations in the values of these universal constants.[27] This discovery acquired the name of 'fine-tuning' of the universe.

Several explanations for the 'fine tuning' observed in the construction of the world and the precision with which nature is seen to operate have been suggested. In one suggestion, this fine tuning was an indication that the universe is meant to be a life-supporting environment, a theory which led to the development of the so-called Anthropic Principle. Other physicists proposed that ours might not be the only universe in existence but be one out of an infinite number of worlds that are simultaneously actual. They called these 'parallel universes.' This hypothesis was called the Multiverse Hypothesis or the Many-Universes Theory.[28] It is possible that all such worlds may be lifeless, except for our own. Consequently, the proponents of this multiverse hypothesis see nothing astonishing in the fine-tuning of our world, since it would be a matter of well-expected chance for a fine-tuned world like ours to exist. The multiverse hypothesis gained publicity with an article by Max Tegmark, which was published in *Scientific American*.[29] However, much of the claim that this multiverse is a reality remains unsubstantiated, and there is no rigorous scientific proof of the existence of such states. George Ellis, the eminent cosmologist, has argued against the actual existence of the multiverse.[30]

The primary source for these speculations is the 'Theory of Chaotic Inflation,' proposed by Andrei Linde. Another source is the Everett interpretation of quantum mechanics, which allows for many other worlds to exist in which

---

27 See for example: Barrow and Tipler, *The Anthropic Cosmological Principle*.

28 For a detailed account about fine tuning, the anthropic principle and the multiverse see: P.C. W. Davies, *The Accidental Universe*. Cambridge University Press, Cambridge, 1982.

29 Max Tegmark, 'Parallel Universes', *Scientific American*, May 2003.

30 Ellis, G.F.R, Does the Multiverse really exist, *Scientific American*, August 2011, 38–43.

all other possible values of a physical measurement are supposed to exist. However, it is important to note that the Everett many-worlds interpretation applies to quantum (microscopic) systems, and not to classical (macroscopic) systems. Furthermore, this is a complicated theoretical problem that cannot be verified or disproved by direct experiments. Another good question which challenges the multiverse hypothesis was raised long ago by Einstein, who asked whether: 'God had any choice when creating this universe.'[31] In the light of the multiverse hypothesis, some physicists say that the answer to this question could be *yes*.[32] The Qur'an suggests that God always has the choice to create other universes with different properties, but such a possibility does not necessarily imply that God has prepared such universes.

In general, while science is firm and strong on the practical side of the story, it is still far from reaching a resolution on the theoretical side. That is why we should not speculate on this issue, since doing so takes us away from reality such that some people may think we cannot maintain an objective view of the world.[33] Instead, we should retain some fixed basic principles and doctrines; some sort of an epistemic paradigm, while finding our way through the relationship between science and religion.

All these, and many other problems, are subject to philosophical and scientific analysis and discussion. Aside from Stephen Hawking's mourning of philosophy, I still feel that there is much room for philosophical thinking and analysis, and this can surely be provided through *Daqīq al-kalām*. Many of the above problems were discussed by the Mutakallimūn of the 9th and 10th centuries, and, today, we can revive those discussions and debates with a wider scope and more accurate vision, enlightened by the power of scientific approach. I would say that *Daqīq al-kalām* can provide a strong basis for developing a consistent and viable philosophy which acknowledges modern science. Additionally, if properly restructured and utilised, *Daqīq al-kalām* can pave the way for a profound reformation in Islamic thought, as well as in other intellectual systems in the modern world. Islamic *kalām* has the potential to reassess our understanding of nature, reality, the world's relationship with God, human destiny, and the purpose of our lives.

31 A. Einstein, quoted in www.humboldt1.com/~gralsto/einstein/quotes.html A. Einstein, quoted in www.humboldt1.com/~gralsto/einstein/quotes.html

32 P.C.W. Davies, 'Multiverse Cosmological Models', *Modern Physics Lett. A* vol. 19 no.10 (2004), pp.727–744.

33 See for example the criticisms of Jim Baggot, *Farewell to Reality*, Pegasus Books; First Edition (2013).

Part II
The Principles of Daqīq al-Kalām

This part is the core of the book. In it, I will explain the basic principles of *Daqīq al-kalām*. These principles have been construed from the books and discourses of early Mutakallimūn, Muʿtazilis and Ashʿaris, and their dialogue on subjects related to natural phenomena and concepts.

At the start, we should pay attention to the necessity of classifying *kalām* into *Daqīq al-Kalām*, or *Laṭīf al-Kalām* and *Jalīl al-kalām*. This classification might not have been very distinctive or accurately descriptive of the actual status of *kalām* practiced over the ages. There was always a mix between the problems and concepts dealing with questions on natural phenomena and those about the divine or humankind. Besides this, it is a fact that some of the concepts that are dealt with in the real natural world were also used in describing entities associated with humans, such as the soul and mind. However, no matter what the terms *Daqīq al-kalām* and *Jalīl al-kalām* meant for the Mutakallimūn over the different ages, I will take the first (*Daqīq*) to mean the trend dealing with problems of natural philosophy, and keep the second to cover all the other problems which deal with divinity and human-related matters.

This classification intends to re-construct *kalām* following a methodology that preserves the aim of patronising it according to the methodology of modern science, whilst segregating those topics of belief, especially the part dealing with divinity, *Jalīl al-kalām*, from topics of natural philosophy. This is so the latter becomes an upper structure for the former, the *Daqīq*, will work as the infrastructure of *Jalīl al-kalām*. This pre-destined plan comes in response for the need to re-establish modern Islamic thought on a scientific basis, thereby making use of the fundamental doctrines laid down originally by *Daqīq al-kalām*, which I find to conform very well with modern science. This would establish tight connections between the Islamic faith and modern systems of thought, specifically, the modern philosophy of natural sciences.

I should stress that this book is not intended to be a historical survey of *kalām*, rather it is meant to be a preliminary thesis through which *Daqīq al-kalām* is reformulated to express Islamic views concerning topics of natural philosophy. Concepts and views of the pioneers of *kalām* will be taken from the available original sources, taking into consideration modern and contemporary assessments wherever available. As I mentioned before, my approach may be selective with regards to this formulation, with the intention of building a basis for the new *kalām* and not just to present historical facts that are of academic interest.

THE PROPOSITION OF KALĀM ACCORDING TO MAIMONIDES

Perhaps Moses ibn Maimon (1138–1204), commonly known as Maimonides, was the first scholar to have clearly identified the basic doctrines of *kalām*. Maimonides was a Jewish Rabbi, physician and philosopher who lived in Cordoba and later left to live in Egypt. Maimonides gained widespread recognition in his medical profession and was appointed court physician to al-Qadi al-Fadil, the chief secretary to Sultan Saladin, then to Saladin himself; after whose death he remained physician to the Ayyubid dynasty.[34]

In his book, *Guide to the Perplexed,* he describes them as "the general premises that the Mutakallimūn laid, irrespective of their views and approaches."[35] Maimonides classified twelve premises (or doctrines), which he called 'propositions.' These are:[36]

- **Proposition I**. All things are composed of atoms.
- **Proposition II**. There is a vacuum (or void).
- **Proposition III**. Time is composed of time-atoms.
- **Proposition IV**. Essence cannot exist without numerous transients.
- **Proposition V**. Each atom is completely furnished with transients (which I will later describe) and cannot exist without them.
- **Proposition VI**. Transients do not continue to exist during two-time atoms.
- **Proposition VII**. The status of habitus is that of their privation, and that the former and latter are all of them existent transients, requiring an efficient cause.
- **Proposition VIII**. All existing creatures consist of essence and transients, and their physical form is a transient.
- **Proposition IX**. No transient can form the substratum for another transient.
- **Proposition X**. The test for the possibility of an imagined object does not consist of its conforming to the existing laws of nature.
- **Proposition XI**. The idea of the infinite is equally inadmissible, whether the infinite be actual, potential, or accidental, *i.e.* there is no difference whether the infinite is formed by a number of co-existing things, or by a series of things. Where one part comes into existence when another has ceased to exist, it is called an accidental infinite. In both cases, the infinite was rejected by the Mutakallimūn as fallacious.

---

34 Julia Bess Frank (1981). "Moses Maimonides: rabbi or medicine." *The Yale Journal of Biology and Medicine*. 54 (1): 79–88.

35 Maimonides, *The Guide to the Perplexed*, p. 195–220.

36 It should be noted that Shlomo Pines has called these propositions "premises."

- **Proposition XII**. The senses mislead and are, in many cases, inefficient. Their perceptions, therefore, cannot form the basis of any law, or yield data for any proof.

Maimonides described these premises in accordance with his own understanding and explained them according to what he thought motivated the Mutakallimūn to set such premises, and their goals in doing so. As I traced back the legacy of *Daqīq al-kalām,* I found that Maimonides' identification is not far from reality, and mostly correct, except that it mixes principal doctrines with some concepts and problems. This is the reason that I think it necessary to re-visit those propositions and extract the most fundamental principles that underlay the philosophy of *kalām* concerning topics of natural philosophy. In this respect, I will identify the principles, name the concepts, and describe the problems separately, so as to enable researchers to have clear vision of the legacy of *kalām* in the light of a new modern vision.

This task is not a simple one; since most of the literature on *kalām* is scattered amongst many books, some of which are incomplete. Additionally, the original writing of the Mutakallimūn, especially the Muʿtazilis, has been largely lost, and the only source for some of their views is the writings of their opponents, which cannot guarantee impartiality or authenticity. The other major problem we face when attempting to trace back the concepts and views of *kalām*, is the use of these same concepts in different contexts. Moreover, some important *kalām* arguments originating from the same Mutakallim are sometimes found in several books about them, or in books which have been written by others. Certainly, we know that the original *kalām* books were mostly in Arabic and, unfortunately, not well-edited.

The majority of the Mutakallimūn generally agreed on the basic doctrines (which I am going to call 'principles') and on their views on natural philosophy problems, but they disagreed on the details. At times, they differed on the application of the principles whilst, at other times, on the related explanations. These principles I have articulated as:

1. Discreteness (atomism)
2. Temporality (creation)
3. Re-Creation
4. Contingency and Indeterminism
5. Integrity of space and time

This identification of the basic principles of *Daqīq al-kalām* enables us to recognise where the views of Mutakallimūn from different groups and factions originate and identify areas of their agreements and disagreements.

In this part of the book, I will try to explain each of the above principles and substantiate my claims using the original sources of *kalām* that I have found in print. Occasionally, however, I will refer to unpublished manuscripts in order to avoid the obscurity of unedited works.

I have avoided detailing the views of the Mutakallimūn and limited my investigation to those areas where I found them agreeing on a specific question. Many of the differences between them are confined to the details of the cases they dealt with. My aim was to discover the consistent system of thought that the Mutakallimūn tried to establish in respect of their belief in the truth of Islam (which was always their starting point). Besides this, it is my aim to construe some commonality that might help reform *kalām* and Islamic jurisprudence to become a modern developing trend of Islamic thought.

My study is limited to the contributions of the pioneering leaders of the Muʿtazilis and Ashʿaris only. No attempt is made to cover the *kalām* of the Maturīdis or the Shīʿa, since I do not believe there is much legacy to be found with them about natural philosophy. Later *kalām*, i.e. after al-Ghazālī, has been mostly avoided also, since I could not find any appreciable novelty in it, although, in most cases, I found that the late Mutakallimūn adopted the same principles of the early *kalām* with regards to natural philosophy. However, this is a separate issue that would need in-depth research and a good deal of analysis of the work done after al-Ghazālī, i.e. during the 12th century and afterwards.

Chapter Three

Atomism

In this chapter, I will present the principle of discreteness (atomism) in *kalām* which involves concepts such as the indivisible part (*aljuz' alladhi la yatajaza'*), the *jawhar al-fard* (plural: *jawāhir*), and the *'araḍ* (plural: *a'rāḍ*), which is the character that describe the properties of things. Then I will try to present the objections posed by two prominent Muslims against *kalām* atomism; one, a theologian and the other, a philosopher, i.e. Ibn Ḥazm and lbn Rushd (Averroes), respectively.

In this study of Islamic atomism, we face a number of concepts which have to be clarified, including the concepts of *jawhar, al-jawhar al-fard, al-'araḍ*, and *al-taḥayyuz*. Others, such as void or vacuum, and *kawūn* (being), and more elaborations on the basic concepts, will be provided throughout the study since these terms are used simultaneously by the philosophers to express different meanings. It should be noted that, sometimes, the Mutakallimūn use the same terminology to mean different things. Shlomo Pines has elaborated on this point in his book, *Studies in Islamic Atomism*, while Harry Wolfson also considered various uses of the same terms in different contexts in his book, *The Philosophy of the Kalām*, and al-Noor Dhanani devoted several pages of his book, *The Physical Theory of Kalām*, to discussing the terms *jawhar* and *'araḍ*. Here, I will present some of the most fundamental terms and designate what they express, as far as I can gather from the study of the original *kalām* books in Arabic.

Jawhar: (plural: *jawāhir*) جوهر was a term used by the Mutakallimūn to express an abstract entity of the *indivisible* part. It is bare of any property, and is assumed to be the smallest constituent of anything that can be envisaged. The word *jawhar* is the Arabisation of an original Persian word meaning 'essence', and not 'substance', as claimed by one author.[37] Also, *jawhar* does not signify the term atom, as claimed by Pretzl.[38] This is because the term atom constitutes the smallest building block of an object, including all its properties (*aʿrāḍ*), while the *jawhar* is only part of the atom. It is the *jawhar,* along with the set of transients (*aʿrāḍ*) which make up the *kalām* atom. This identification is very important in order to designate the concepts properly. Some Western authors have used the word 'substance' to express the meaning of the term *jawhar* in English, and since the term *substance* is a well-defined term in Greek philosophy, it may lead to serious confusion with the corresponding Greek concept.

Al-Jawhar al-fard: (plural: *al-jawāhir al-farda*) الجوهرis a term used to express the affirmed concept of the singularity of the *jawhar*. This should not be understood to have the same meaning of *al-jawhar al-waḥid* (the single *jawhar*) since the term *al-fard* is meant to be *singular* in all respect thereby stressing its abstractness. Note that the word *al-waḥid* is not a specified term but a normal word meaning one or single. This becomes clear when we read these expressions as used by pioneers of *kalām* in their context.

ʿAraḍ: (plural: *aʿrāḍ*) عرضpoints to the entities, characters or meanings that are possibly accepted by the *jawhar.* It describes its attributes and additive properties, rather than any intrinsic properties. Most authors use the word *accident* to express the meaning of the term *ʿaraḍ* in English, and, while this might not reflect an accurate meaning, I consider it acceptable for use to describe the attributes of a body rather than the *jawhar*. However, it should not be confused with a similar term in Greek philosophy which is used in a similar context. The term *ʿaraḍ* points to a *transient* associated with the *jawhar*, whereas the term *accident* (in Greek philosophy) is associated with bodies only.

Juz' (plural: *ajzā'*): جزءthis term might seem equivocal but certainly in Arabic it means *the part.* The Mutakallimūn used it to mean the smallest possible part of the body. This allows me to understand that it describes the *jawhar* together with the possible *aʿrāḍ* that it may carry. This will then represent the *atom.*

Taḥayyuz: تحيّز is a term used by the Mutakallimūn to express, basically, the existence of the *jawhar*. Sometimes this term is explained to mean the

37 Dhanani, al-Noor, *The Physical Theory of kalām*, p. 55.

38 Pertzl, O. (1931), Die fruhislamische Atomleher, Der Islam 19, 117–130.

occupation of space,[39] but this is incorrect, since if it was meant to occupy a space, the *jawhar* should have extended dimensions. This could be accepted for the body but not for the *jawhar*. Perhaps it is more suitable to use the word *occupancy* in English to express the meaning of *taḥayyuz*. The *mutaḥayyiz* is the occupant.

The most prominent leader who, it is claimed, suggested the notion of discreteness was Abū al-Hudhayl al-ʿAllāf. He was followed by Muʿammar Ibn ʿAbbād al-Sullamī and the rest of the Muʿtazilis. The Ashʿaris also approved of the notion of discreteness and have elaborated it further to use it in their arguments in *Jalīl al-kalām*. The Maturīdis and Shīʿa are also known to have endorsed the notion of atomism.

CONTINUITY AND DISCRETENESS

The most fundamental assumption of *kalām*, discreteness, is the claim that all objects and entities in nature can be subdivided into smaller and smaller parts until we reach a minimal part that can no longer be subdivided. This indivisible part is called the *juzʾ al-ladhī la yatajazaʾ* or the *juzʾ* for short. Al-Sharīf al-Jurjānī (1339–1414), who collected the terms and expressions used in *kalām* in his book, *al-Taʾrīfāt*, defines this minimal part (*aljuzʾ allathi la yatajazaʾ*) as "an essence that does not accept any further division, neither in imagination nor in reality. The bodies are formed out of these parts by composition."[40]

Al-Bāqillānī says, "All created objects can be classified into three types: a composed body, a single *jawhar*, and the *ʿaraḍ* (attribute), which is acquired by the bodies and the *jawāhir*."[41] This means that anything in the world is, *theoretically*, one of these three types. I emphasise *theoretically* here because we can actually only see the bodies with their attributes (*aʿrāḍ*) and not their *jawāhir*, for these cannot be found as stand-alone objects. Al-Jūwaynī (d. 1123) points to this abstractness when conceptualising the *jawhar* by saying that, "Muslims agreed that bodies can be finitely subdivided into parts and each of those non-divisible parts has no boundaries or a defined extension and this is what geometers have understood and they have described the part by a point and said that the point is non-divisible."[42] According to this understanding of the Mutakallimūn, the indivisible part is the smallest element composing the body. Most Greek philosophers, including Aristotle, believed that things are infinitely divisible.

Here, I will only present those views which have gained the acknowledgment of most of the Mutakallimūn, irrespective of their subscription to the

39 Dhanani, physical, p. 62.
40 Al-Jurjānī, *al-Taʿrīfāt, al-Juzʾ*.
41 Al-Bāqillānī, Tamhīd, 37.
42 Al-Jūwaynī, Al-Shāmil, p. 158.

different schools of thought. For this reason, the reader should be aware that the different views of the Mutakallimūn, in respect of the details of their principles, will not be discussed.

Arguments Supporting Finite Divisibility

In this section I am going to briefly give the main arguments of the Mutakallimūn in support of their assumption of finite divisibility. I will not present the details of their arguments here. Much evidence was presented by al-Shahrustānī in his book, *Nihayat al-Iqdām fī 'Ilm al-Kalām.*

Al-Ashʿari reports that Abū al-Hudhayl al-ʿAllāf claimed that, "God has the power to disintegrate the body till it becomes a non-divisible part, and than this part has no length, width or depth."[43] However, al-Jūwaynī proposed a more rational argument, saying, "we argue against them (the philosophers) that if one looks at a small body and a large one, then we know necessarily that one of them is larger the other. Then if we are sure of this, the differences between the large and the small should either be attributed to the larger number of parts [contained in the large body] or not. If it is attributed to the number of parts, then the finiteness of the body is demonstrated; since if the two bodies were to be infinitely divisible then there can be no arguments that one could contain parts more than the other."[44] In this context comes their statement that, "whatever is finitely bounded should be finite and whatever is unbounded could be infinite." Abū Saʿīd al-Sīrafī, a well-known Muʿtazilī, writes, "the *jawhar* is every part which enjoys *taḥayyuz* (occupancy)."[45] This means, he continues, "that it should not be understood that the *jawhar* is occupying that place,"[46] suggesting that the *jawhar* is not specifically in that place. By this, I understand that *taḥayyuz* does not mean *occupation of space.* A similar opinion is presented by al-Jūwaynī. This will be discussed further later.

Later in this chapter, I will present some critiques that have been put forward by the famous theologian, Ibn Ḥazm, and the famous Muslim philosopher, Ibn Rushd, about the concept of the *jawhar* and the assumption of the existence of an indivisible part. This may help clarify the position of these two important scholars on *kalām* atomism. In this respect, it is important to note the different motivations and schemes of thought that each of these scholars followed. Otherwise, the subject could be widened to discuss the different views of later Mutakallimūn about atomism, specifically the views of al-Fakhr al-Rāzī and the critiques of Ibn Sīnā and his influence on the later

43 Al-Ashʿari, Maqālāt, p. 314.
44 Al-Jūwaynī, Shāmil, 146.
45 Al-Sīrafī, al-Ghunyah, 1/50.
46 Ibid.

Mutakallimūn. However, such a detailed analysis is out of the scope of this book and can be found with some other authors.[47]

THE TRANSIENTS (A'RĀḌ)

The *a'rāḍ* (plural of *'araḍ*), which is mistakenly called 'accidents', are the attached companions of the jawāhir. No *jawhar* can be sustained without the presence of its *a'rāḍ*, for at least one *'araḍ* must be present at any moment for the *jawhar* to be identifiable. However, most *a'rāḍ* do not endure, according to the Mu'tazilis, while according to the Ash'aris, none of the *a'rāḍ* endure. This is the reason for calling them *transients*, as they are accidental attributes subsisted by the *jawhar*.

Al-Jurjānī defines the *'araḍ* as, "the existent which is in need for a substrate (*maḥil*), that is a place to hold it. Examples are: colours, flavours, the taste and touch, and other characters that are impossible to endure."[48] Of course, al-Jurjānī presented the views of the Ash'aris and the early Mutakallimūn. He also declared that, "the *'araḍ* is a transient that appears in the *jawhar* like colour, flavour, taste, touch and others, which do not endure once existed."[49] This means that we can consider the material properties of things as transient peculiarities that are always in need of being held by a substrate (which is the *jawhar*). Such properties are not durable, as they are thought to be subject to change in the next moment of their existence. This requirement has been reflected in the *kalām* notion of re-creation.

The Mutakallimūn generalised the concept of *'araḍ* to include power, will, life, potency and death. This kind of generalisation reflects the primitive thinking of the early Mutakallimūn and their crude usage of the term in more than one context. Such a use, I feel, had the negative effect of causing confusion and hindering the proper development of *kalām* by mixing natural topics with human and divine (wilful) actions. This situation could cause the reader of al-Ash'ari's *Maqālāt* some confusion with so many diverse views being aired. Since the early Mutakallimūn held a clear distinction between conscious and wilful entities, and inanimate matter, they should have differentiated between the use of the attributes (*a'rāḍ*) of each of these, but we do not see this in their discourse.

---

47 See for example, Pines, *Atomism* pp. 41–107; Dhanani, The impact of Ibn Sīnā's critique of atomism on subsequent *kalām* discussions of atomism, *Arabic Science and Philosophy*, vol.25, 79–104 (2015).

48 Al-Jurjānī, al-Ta'rīfāt: *'arāḍ*.

49 Ibid.

On the endurance of the *aʿrāḍ*, al-Ashʿari presented different views of the Mutakallimūn, saying:[50]

1. Some said, "They are called this because they [are transient phenomena which] occur *(yaʿtariḍu)* in bodies and subsist in them *(taqūmu bihā)*. They denied the existence of transients which are not in any substrate *(la fi makān)* and transients which are not created in bodies."
2. Some others said, "It is possible for transients which are not in bodies and for events (*ḥawādith*) which are not in any substrate to exist like time (*waqt*), God's volition *(irāda)*, the endurance *(baqāʾ)*, and ceasing to exist *(fanāʾ)*, and the act of creation (*khalq*), which consists of the acts of speech and willing by God." This is the doctrine of Abū al-Hudhayl, according to Ashʿari.
3. Others said, "The *aʿrāḍ* are called transient because they do not endure (*lā labtha lahā*)."
4. Some said, "The *ʿaraḍ* is called transient because it is not self-subsistent (*lā yaqūmu bi-nafsihi*) nor is it a kind of thing (*jins*) which is self-subsistent."
5. Some said, "The entities which subsist in bodies are called transients as a result of a convention adopted by the *Mutakallimūn*. If someone were to deny the use of this term [for the transient], we could neither present an argument (*ḥujja*) to him based on the Qur'an, nor the Tradition of the Prophet, nor the consensus of the community, nor of the lexicographers. This is the view of (*ahl al-naẓar*), one of whom is Jaʿfar Ibn Harb."
6. ʿAbd Allah Ibn Kullāb called the entities which subsist in bodies transients, things (*ashyāʾ*), and also described them as attributes (*ṣifat*).

Another prominent leader of the Ashʿaris, Abū Bakr al-Bāqillānī, says, "ʿ*arāḍ* are those which do not endure, and which emerge in the *jawāhir* and bodies, and cease to exist in the next moment of their existence."[51] Basically, the word, *ʿarāḍ*, in Arabic points to a transient that does not endure. Al-Jūwaynī says of this meaning, "originally, linguistically, the *ʿaraḍ* is a transient in existence that does not endure, whether it was a body or [*jawhar*], this is the reason for calling the cloud *ʿaraḍ*."[52] Al-Bāqillānī establishes that the presence of *ʿaraḍ* is through the existence of motion, "The proof for [the existence of] *ʿaraḍ* is the movement of a body after it was at rest and getting to rest after it was in motion. Such a status can be either self-produced or by a cause. If [motion] is self-produced, it would never stop. The fact that it goes to rest after motion is evidence that its motion was caused."[53]

50 Maqālāt, 369–370.
51 Tamhīd, p. 38.
52 Shāmīl, 169.
53 Tamhīd, 38.

It should be noted that, according to the understanding of most of the Mutakallimūn, an event that happens once, naturally and intrinsically, and is unaffected by any external agency, should then go on happening. This has been used on many occasions as an argument against the claims of naturally occurring phenomena, such as plant growth or getting drunk by drinking wine. This persistent action is attributed to the absence of will. This issue will be discussed in Chapter Six below and elaborated in Part III when dealing with the subject of causality.

PROPERTIES OF THE JAWHAR

From what we see in the writings of the Muʿtazilis, the *jawhar al-fard* is an entity without magnitude. Al-Ashʿari reports that Abū al-Hudhayl claimed that the *jawhar* has no volume, since it "has no length, no width, or height", nor does it have an extension or sides "right, left, back, face, upper or lower part."[54] Consequently, I would suggest that, according to Abū al-Hudhayl, the *jawhar* is an abstract entity that does not occupy space but exhibits its existence by having some sort of occupancy, which is either the apportioning of space (*taqdīr al-makān*), or a tendency to join another *jawhar*, thereby forming an extension.

Indeed, this is what the Mutakallimūn thought when they said that the minimum number of parts forming an extension (a line) is two. If we have four parts, then we can have two-dimensional surface: namely a plane. And, if we have eight parts, we can construct a volume, say a cube. In this respect, it was reported that Abū al-Hudhayl found that a body can be formed out of at least six parts: one to be the right, a second for the left, with the third and fourth forming the face and the back, and the other two forming the top and bottom of the body. It is similar to the six faces of the cube being replaced by six tiny spheres all touching each other. This what Ibn Matawayh confirms in his book, *al-Tadhkira*, "The way we get a construction out of these parts is to have two of them making a length called a line… then as we put two other parts to form the width we obtain a surface, like a plane, then if we put on top of these four [parts] other four parts then we obtain along with the length and the width, the height; a body."[55]

In *Maqālāt* by al-Ashʿari, we also read that the leaders of the early Mutakallimūn asserted that the jawāhir can never be a body, "Every *jawhar* is not a body. It is impossible for a single *jawhar*, which is indivisible to be a body because this is that which is long, wide, and deep. The single *jawhar* is not like

54 Maqālāt, 302.
55 Tadhkirah, 71.

this. This is the doctrine of Abū al-Hudhayl, and Muʿammar, [Abū Alī] al-Jubbā'ī followed this doctrine." [56] The view that the *jawhar* has a magnitude of its own was affirmed by the famous Ashʿari, al-Jūwaynī, "The singular *jawhar* has a fixed share (*ḥaḍḍun thābit*) of area that is not restricted to adding another [jawāhir] to it, and has a magnitude, but its magnitude has no parts, and the *jawhar* is valued by the *jawhar*."[57] Al-Naysābūrī reports that Abū Hashim al-Jubbā'ī agreed with this. [58] According to the forgoing statements, the *jawhar* is a point-like entity that can be identified with its occupancy and realised once it joins at least one more *jawhar* to form the shortest line. The *jawhar* on its own cannot be identified unless it is associated with at least one transient attribute, an *ʿaraḍ*.

Taḥayyuz (Occupancy)

This is an intrinsic property of the *jawhar*. It is not something that is added to it, as is the case with the body. According, to al-Jurjānī, the *ḥayyīz* is "the envisaged emptiness which is occupied by an extended object, like the body, or un-extended object, like the *jawhar*."[59] This statement implies that the *jawhar* is not an extended object and, consequently, its occupation of space should be essentially different from the occupation of space by a body. For this reason, I will use the word 'place' to express the meaning intended for what the *jawhar* would occupy and 'occupation' to express the term *taḥayyuz*. Al-Jūwaynī has admitted that *taḥayyuz* is an equivocal term. So, he says that this term is among those where the Mutakallimūn differ, and then tries to describe the concept in subtle phrases saying, "If someone asks: What, then, is *taḥayyuz*? Some of the leading authorities [of *kalām*] have said: It is the apportioning of space (*taqdīr al-makān*). They do not, by this, mean that when God creates a singular *jawhar*, its *taḥayyuz* is the apportioning of the place which belongs to it, for this [entails that *taḥayyuz*] awaits [the appearance of] an existent. But *taḥayyuz* is a real thing and not something which is anticipated. Its meaning, rather, is that it is a place (*makān*) that belongs to a *jawhar* which has apportioned [this place]." Then he adds that, "The best statement which can be said about the *taḥayyuz* is the *mutaḥayyīz* itself, and the meaning of the *mutaḥayyīz* has been mentioned before. (Pointing to the statement mentioned before in which he says: the *mutaḥayyīz* is the existent by which no other of its like can be found). Also, it would be possible to attribute the *ḥayyiz* to the *jawhar* as it would be possible to add existence to it.[60] In one way or another, al-Jūwaynī seemed to

56 Maqalāt, 301.
57 Shāmīl, 159.
58 Masa'il, 58.
59 Al-Taʿrīfat: ḥayyīz.
60 Dhanani translates the last sentence of al-Jūwaynī as "The clearest statement that can be

experience some difficulty in expressing the abstract nature of the *jawhar*. Occupation of place is one of the most effective concepts that may actualise the *jawhar* to become a corporeal object, a body. This is what al-Jūwaynī and the other Mutakallimūn, especially the Muʿtazilis, did not like.

The above understanding is supported by Ibn Matawayh, "The truth of *jawhar* is what has ḥayyīz (place) upon its existence, and the *mutaḥayyiz* (the occupant) is that [thing] specified by a state of being able to grow by being combined with others of its kind, or occupy a differential part (*qadar*) of the place, or what is apportioning of space (*taqdīr makan*), so as to acquire that place, or prevent its alike from acquiring that space."[61] Again, the fact that a *jawhar* (accompanied by its transients) might combine with another to form a segment would certainly require place since, in this case, the combination of the two *jawhar* will need space to occupy. It was this requirement that made the Mutakallimūn find it necessary to apportion the *jawhar* a certain amount of space, or *taḥayyuz*. Otherwise, place would have to be created once two *jawhar* were combined. Such a creation has no mechanism in *Daqīq* al-*kalām* and this is the reason for the concept of *taḥayyuz* to be sufficient enough to provide the needed place for an extended combination.

Taḥayyuz does not denote the occupation of a place in the same way as a box of matches does. If the *taḥayyuz* was meant to mean the 'occupation of space' in the way Dahanain presented it, then why should al-Jūwaynī, Ibn Matawayh and the other leaders of the Mutakallimūn be puzzled as to its meaning and find it a difficult task to communicate the wording? Certainly, to envisage such an abstract concept needs all such involvement of terms, expressions, and meanings.

THE VOID (*AL-KHALĀʾ*)

The concept of void in *kalām* is closely related to the concept of non-being (*maʿdūm*: the non-existing, the nullified). This topic is important in *kalām*, as well as in modern physics, for several reasons. First, for its intimate relation with the meaning of the existent, second, for its connection with the mechanism of creation and re-creation and, third, for its implications in respect to eternity and the possibility that the world originated from an eternal thing which had existed for an infinite extension of time.

---

made regarding *taḥayyuz* is that it is, in itself, space-occupying (innahu mutaḥayyizun bi-nafsihi) and the meaning of an object which occupies space has been presented above. Moreover, the relationship of the atom to *taḥayyuz* is not difficult to conceive just as the relationship of existence to the atom is not difficult to conceive." Dhanānī, physical, p.65.

61 Tadhkirah, 47.

As defined by al-Jurjānī, a void is "the envisaged place which is unoccupied by a body."[62] It is known that all Muslim philosophers, except Abū Bakr al-Rāzī, negated the existence of the void, while all the Mutakallimūn, except al-Ka'bī, affirmed it. The reason for this was the need to assume the existence of voids so as to allow for the motion of the indivisible parts i.e. atoms. In using the term void, the Mutakallimūn meant the empty space that contains nothing. Such an empty space, they assumed, did not necessarily exist simply for the presence of a body. In modern terms, we call this a "vacuum." Historically, we learn that the Mutakallimūn were divided in their views about the reality of the void. Some said that the void actually existed, while others claimed it was only an envisaged entity that had no realisation. The latter argued that should the void be realised as an extension, it would be a body and, on being a body, it would be observable and, since no void is observed, it can only exist in our mind.

Those who affirmed the existence of voids tried to demonstrate that they exist. They suggested experiments, the results of which could not be interpreted accurately except by assuming the existence of a void. Al-Fārābī, the philosopher, reports that the Mutakallimūn took a cup and immersed it upside down completely in water. They then removed it, only to find that the water had not touched the top inside surface of the cup. This, they said, meant that the bottom of the cup had been full of air when it was immersed. Then they brought a glass jar with a wide bottom and narrow opening. They then sucked the air out of it, closing the opening with a finger and immersing it upside down into the water. Upon doing this, they found that water filled all the space inside the jar, meaning that water had replaced all the air. Such a replacement affirmed for them the existence of an empty space, which was the void. This story is very interesting as it shows that the Mutakallimūn used experimental demonstrations to argue in defence of their views. Jaḥiẓ tells us that it was known that al-Naẓẓam resorted to experimentation,[63] while Abū Rashid al-Naysābūri and Ibn Matawayh have given several examples of experiments that confirm the reality of the void. Most of the philosophers denied the existence of voids because they could not accept any extension of dimensions without the presence of bodies (or some form of matter). If the void has any dimensions, then it would be a body, and a body occupies a space.

As we see, the argument put forward by the Mutakallimūn for the need of the void—as a necessary space through which motion of the jawāhir can be achieved—is surely acceptable as motion is not possible through a filled space. On the other hand, the philosopher finds no way to acknowledge the rational

62 Al-Ta'rīfāt: *al-Khala*

63 Al-Jaḥiẓ, Ḥaywan.

existence of the void since air, for example, will naturally move in to fill the void.

Our modern understanding of matter and motion allows for empty space to exist. In fact, most of the universe, and even the space within the atoms, which compose matter, is a vast void. The solar system, and its inter-stellar medium, is mostly a rarefied place that contains very few atoms or dust particles per cubic metre. However, we should also notice that the ideal void, meaning the vacuum, as understood by modern physics, is not the same as the classical nothing. Rather, the modern understanding of the physical vacuum is something else. Vacuum (void) in modern physics is thought to be full of "virtual" particle-antiparticle pairs. This brings us to discuss the concept of "nothingness."

The Nothing and the Non-being

Historically, there has been strong debate, and even conflict, between the Muʿtazilis and the Ashʿaris in respect of the non-being (*maʿdūm*) and the nothing i.e. the vacuum. This arose in discussions regarding the problem of creation. The Muʿtazilis described the non-being as composed of jawāhir without occupancy. Accordingly, the vacuum, in this description, was not an absolute nothing. In contrast, the Ashʿaris described the non-being as an absolute nothing. Al-Jūwaynī says of this, "people of truth, (the Ashʿaris), has adopted that the reality of a 'thing' is to be existing so that every existent is a 'thing' and any 'thing' should be existing, and whatever cannot be described as a 'thing' cannot be called existing."[64] He goes on to say, "the non-being is null in all aspects, and the meaning of being *known* is to indicate that it does *exist.*"[65]

The Muʿtazilis adopted the understanding that the non-being is a thing because, as al-Jūwaynī reports, "the truth of the thing is to be known, consequently, they said that every non-existent is a thing, and they claimed that it is a *jawhar* in nullified state. However, they negated some properties and did not describe the non-existent *jawhar* with the property of occupancy (*taḥayyuz*) despite the fact that the occupancy is originally an intrinsic property [of the real *jawhar*] according to them."[66] This is an important statement and should be read in Arabic with much care, in order to fully grasp its meaning. Here, I have tried to present it with the best possible expression in my capacity.

The importance of this problem comes from the fact that by saying a nullified *jawhar* exists, we may imply that the *nothing* is implicitly *something*. In this case the *nothing* might be understood to have an eternal existence, which

64 Shāmīl, 124.
65 Ibid.
66 Ibid.

contradicts the Islamic belief of having God (Allah) as the only eternal existent. This was, and still is, one of the major problems we face when confronting religious beliefs with rational thinking. The Ashʿaris refused to accept the claim that the nothing might be something. It is interesting to note that this deductive conclusion that the vacuum, or non-being, could be called a *thing* has been obtained following the rules of the *kalām* deduction rules.

On describing the Muʿtazili position on the composition of the non-being, al-Naysabūrī reports that Abū Alī al-Jubbā'ī and his son, Abū Hashim al-Jubbā'ī, believed that the non-being is composed of jawāhir, so he says, "know that what the two Shaykhs, Abū Alī and Abū Hashim, believed is that the *jawhar* is a *jawhar* even when non-being."[67] Al-Naysabūrī also says, "and this is what Shaykh Abū Abdullah (Qāḍī ʿAbdul-Jabbār) has mentioned about them too." Then he adds, "and it may appear from what he says that the character of occupancy (*taḥayyuz*) may happen for the non-being, but the condition that occupancy would deprive a similar [*jawhar*] from being where it is, prevents the non-being from obtaining the character of occupancy unless it exists, so he takes the existence as a condition for the feature of occupancy."[68] However, it seems that the Muʿtazilis of Baghdad negated this description of the Basrian Muʿtazilis that the non-being could be composed of jawāhir. Al-Naysabūrī says, "our master, Abū al-Qasim [al-Kaʿbī], believed that the non-being cannot be described as *jawhar* and not a transient (*ʿarāḍ*) and he refrained from calling it other than a thing (*shay*')."[69]

Al-Naysabūrī continued to question the non-being from several angles to prove what his master, al-Kaʿbī, the Baghdadī Muʿtazili, adopted with regards to negating the vacuum being composed of non-being jawāhir, despite acknowledging the non-being as a thing. However, al-Jūwaynī, who was an exponent of the Ashʿari, denied that al-Kaʿbī acknowledged the non-being as a 'thing' and said that, "al-Kaʿbī and his follower, amongst the Muʿtazilis of Baghdad, agreed with the people of the truth [the Ashʿaris] in saying that the non-being is nothing and that it is absolute null."[70] Thus, in summary, I can confidently say that the Ashʿaris agreed that the non-being is a 'nothing', the Basrian Muʿtazilis claimed that the non-being was composed of un-occupational jawāhir (*jawāhir ghayr mutaḥayyizah*), and that the Muʿtazilis of Baghdad, headed by al-Kaʿbī, while negating the claim that the non-being is some sort of jawāhir, agreed that it could be 'something,' perhaps reckoning that nothing can be produced from nothing. In this sense, the process of creation becomes an unfolding existence out of the non-being, a notion that is echoed

67 Al-Naysābūrī, Masail, 37.
68 Ibid.
69 Ibid.
70 Shāmīl, 125.

strongly in the concept of the physical vacuum, as expounded in theories of modern physics. This will be explained shortly.

ʿAbd al-Jabbār, in his book, *Al-Muḥit bil taklīf,* considered the question of whether the non-being can be numerated. This question he raised in the context of describing the power of Allah and His knowledge. He says, "If the abilities of Allah were to be unlimited, and if the subjects of his ability cover the non-beings, then His knowledge should be unlimited if He is to be acknowledged as omniscient. This shows that there is no method available to prove the non-being, except by resorting to knowing the state of the omnipotent, and that this, superficially, indicates His unlimited power, and confirms what we just said. Accordingly, we should understand the verse (and he counted everything in numbers) to apply only on those existents which can be counted not to the non-beings which are unlimited."[71] In this statement, ʿAbd al-Jabbār was expressing his belief that non-beings are infinite, and that what we call a vacuum is a continuous medium. He signifies here that, although the vacuum is a 'thing', it can only be understood if we consider it continuous and uncountable.

This is consistent with the Muʿtazilis confirmation that the jawāhir of the non-being have no occupancy. This is expressed by the fact that, as long as the non-being has not been realised ontologically, it does not exist, but that once it is converted into a being (created), then it will change from its epistemological state (being known) to a state of a real being that ontologically exists. This means that the world of non-beings (the vacuum) is an ocean of all contingencies which can be created by the will and power of Allah. It might be of interest to know that the Iraqi philosopher, Hussām al-Alusī, has discussed in detail the question of divine attributes and their relationship with the idea of non-beings as something rather than nothing. This discussion can be found in his Ph.D. thesis, which was submitted to Cambridge University in 1965 under the title: *The Problem of Creation in Islamic Thought, Qur'an, Hadīth, Commentaries and Kalām.*[72]

The Non-being in Modern Physics

I hope I can convey my surprise and astonishment when I speak of the richness of the notions that the Muʿtazilis held regarding the non-being. As a theoretical physicist specialising in quantum field theory and general relativity, I can fully appreciate the scientific value of these notions. In our study of the creation of the universe, according to the available theories of modem physics, we

71 ʿAbd al-Jabbār, Muḥīt 117.

72 Al-Alūsī, The problem of creation in Islamic Thought Qur'an, Ḥadīth, Commentaries and *kalām*, Ph.D. dissertation, Cambridge University, 1965. pp. 191–223.

find no alternative but to make use of the Heisenberg Uncertainty Principle. In one version of this principle, we see the vacuum (the non-being, the nothing) as a boiling state of particle-antiparticle pairs. These are created such that they only endure for an immeasurably short period. This is because the pairs are annihilated in the next instant of creation and, thus, cannot be detected by any instrument. These particle-antiparticle pairs are called *virtual* states since they cannot be measured directly, i.e. they do not exist in the physical meaning of existence. To exist physically, an object has to be measurable. If deemed unmeasurable, it is considered as being non-physical. For example, love is a non-physical entity, despite the fact that it may be felt in conjunction with some physical and intimate practices. Reflection and belief are non-physical entities, despite their involving chemical and biological activities in the brain. Again, the fact that these are non-measurable entities makes us consider them non-physical.

Even though virtual particles are considered, ontologically, as non-beings, their effects (i.e. their virtual state) can solve some serious problems in nuclear physics. For example, the nuclear force between protons and the neutrons inside the nucleus can only be understood if we assume that such virtual states exist. The vacuum is assumed to be composed of virtual pairs of all sorts of particles. The question is how such virtual states can be converted into real states?

Convention suggests that an external force, gravity, for example, is needed to convert the vacuum (the non-being) into something (a being). What happens is that gravity causes time to dilate (be prolonged) and, consequently, the duration of time that such virtual states can exist, is extended. If these states remain viable for long enough, they become measurable, and once they become measurable, they are real. In this way, they are affected by their physical environment, interacting with it to become part of the real world and no longer of the virtual world. But from where does the excess energy come to convert these virtual states into real states? It comes from the external field: e.g., gravity supplies the needed energy.

This sort of explanation became possible only after we had the 20th century revolutionary theories of relativity and quantum mechanics. The calculations of such processes, which explain the creation of the universe, are very complicated and involve high mathematics. However, the realisation of the process of creation, according to this explanation, seems to be simple. In the case of the creation of the universe from nothing (*ex nihilo*), we can imagine that the vacuum stands as the state of zero total energy. This state is composed of many virtual particle-antiparticle pairs. We cannot catch such pairs because they live for very short periods of time. Now, because gravity comes in the form of a curved spacetime, it will prolong the duration of such pairs. In doing so, it absorbs the negative energy (-1, i.e. the antiparticle) of the zero (the vacuum), whilst leaving the positive (+1, i.e. the particle) energy of the vacuum, which will appear then like a real particle with positive energy. Meanwhile, the curved spacetime becomes flattened as spatial curvature is weakened during the creation of the 'real' stuff. This causes spacetime to expand and become flattened continuously.

I have studied this interplay between the curvature of spacetime and the vacuum in my professional scientific research.[73] Incidentally, Paul Dirac, the great physicist, discovered that the physical vacuum must be composed of negative energy states that are full. These energy states form a continuum called the *Dirac Sea*. Whenever a negative energy state becomes empty it has to appear in the real world as an anti-particle. For example, the empty negative energy electron state (originally conceived of as a hole) will appear as a positron (a positively charged electron) in our world. Therefore, real particles were holes in Dirac's negative energy states. If you compare all the particles of the real world to the holes of the Dirac Sea, the net result is zero. This what Lawrence Krauss meant when he talked about "a universe from nothing."[74] The discovery above might help us find the scientific significance of the Muʿtazilis' notion that the non-being is something rather than nothing.

TYPES OF JAWĀHIR

Al-Ashʿari reported that al-Jubbāʾī considered all jawāhir to be of the same type. This is what most of the Mutakallimūn agreed upon. However, al-Naysābūrī reported that al-Kaʿbī did not fully agree with this, saying, "Our masters [al-Jubbāʾī] believed that all the jawāhir are of the same type and our shaykh, Abū al-Qasim al-Balkhī (al-Kaʿbī), believed that the jawāhir could be different as well as being similar."[75] On the other hand, al-Jūwaynī reported that, "all Ashʿaris agreed that the jawāhir are of the same type and the Muʿtazilis have agreed to this position, except al-Naẓẓām who did not consider the jawāhir alike unless they were alike in their common transients."[76]

The important implication of the jawāhir being the same type supports the belief that they were abstract entities that take on their ontological character only after becoming associated with a transient. Accordingly, the differences between objects or bodies comes from the different kind of attributes

---

73 Altaie, M. B., and Dowker, J. S., *Spinor Fields in an Einstein Universe: Finite temperature corrections*, Physical Review D18, 3557, (1978).

74 Krauss, Lawrence M. *A Universe from Nothing: why there is Something rather than nothing.* New York: Free Press, 2012.

75 Al-Naysabūrī, *Masail* p. 29.

76 Al-Jūwaynī, *Shāmīl* p. 153.

subsisting the jawāhir of that object. All the apparent properties of objects are due to the different attributes they acquire, otherwise the jawāhir would be the same. Therefore, if we imagine that the jawāhir are meant to be some sort of elementary particles, in the modern understanding of the term, then they should be a primitive sort of particle and not the ordinary, elementary ones. Jawāhir, actually, are a kind of unifying state of all the known particles in the universe that can be recognised only after acquiring transients, such as mass, charge, and spin. Such a philosophical vision can only be found in *kalām*!

This is the far-reaching vision of the Mutakallimūn indeed, and what makes the *kalām* concept of atomism fundamentally different from other concepts of philosophical atomism. This kind of vision confirms that the Arab-Islamic mind was able to deal with abstract concepts as well as corporeal objects; contrary to some authors' claims that the Arabic-Islamic mind is only descriptive.

THE WEIGHT OF THE JAWĀHIR

As for the weight of the jawāhir, Ibn Matawayh says that the *jawhar* on its own has no weight. But it seems there were some disagreements between Abū Alī al-Jubbā'ī and his son, Abū Hashim, on this issue. Ibn Matawayh reports, "and it (the *jawhar*) has no share of weight, but the reference for weight is some sort of entity (*ma'nā*) in it. This is the opinion of Abū Hashim. Abū Alī believed that weight is an intrinsic property of the jawāhir, so he has attributed some share for it, but this question is better understood in terms of the adherences (*I'timādat*)."[77] I find the last sentence in this quotation very important, since attributing weight to the adherence (*I'timādat*) points to weight being not just a non-intrinsic property, but rather a property that might be introduced through the interaction of the object with the environment. This is again something that one may reflect upon in the light of our physical understanding of weight.

It is notable that a fundamental aspect of *kalām* vision, with regards to the abstract jawāhir, is that, in many cases, the single *jawhar* acquires its ontological properties through interaction with its surroundings. In the above statement, weight is acquired through an interaction with an *I'timādat*, which is always related to something else, for example, gravity. As they say: a stone thrown upwards goes upwards because of the *enforced adherences* (*I'timādat mujtalaba*), i.e. the pushing force, and will fall back down due to the *necessary adherences* (*I'timādat lāzima*), i.e. the forces of gravity. In another instance, we find the *jawhar* becoming an object once it is in the state of being combined

77 Tadhkirah 183.

with another, in this case, a body. Al-Ash'ari reports that some of the Baghdadi Mu'tazilis hold that, "If the indivisible part is combined to another then each of them is a body as long as they are in combination; if they get separated, none of them is a body."[78] In this way, they are just like quarks!

The jawāhir do not merge but can neighbour each other, and no two *jawhar* can fuse into one. Al-Jūwaynī says, "what the people of truth have agreed upon is that the jawāhir do not merge together and it is not possible for a *jawhar* to be fused into another, though if it is crudely said that the jawāhir are merged and mixed, then we mean that it is neighboured."[79] In respect of the endurance of jawāhir, most of the Mutakallimūn agreed that these are immutable, unlike the transient attributes (*a'rāḍ*). Al-Jūwaynī reports that, "the *jawhar* is fixed, unchanging, but al-Naẓẓām said that it is renewable."[80] The point to make here is that, although jawāhir are thought to be intrinsically immutable, as their existence is always associated with the mutable transient attributes (which they have to acquire), it would be feasible to think that the jawāhir are *practically* mutable too. However, this can be accepted on the epistemological level but not on an ontological one.

AL-NAẒẒAM'S VIEWS ABOUT ATOMISM

All the available sources of *kalām* agree that Ibrahim Al-Naẓẓām denied the existence of the invisible part and believed in the infinite divisibility of things. Al-Ash'ari reported that al-Naẓẓām wrote a book about the part (*aljuz'*), which indicates his interest in this topic. Al-Naẓẓām is considered to subscribe to the Mu'tazilis' line of thought because he believed in the five theological doctrines of the Mu'tazilis. He was a student of Waṣil Ibn 'Atta' and the mentor of Abū 'Uthmān al-Jaḥiẓ (776–868), the famous writer. Al-Jaḥiẓ tells us many stories about al-Naẓẓām in his books, showing that he was very much interested in doing experiments.[81] Al-Ash'ari also reports al-Naẓẓām as saying that, "the body is the long, the wide, the high, that its part are not numerable and that there is no half unless it can be halved and that there is no part unless it can be parted."[82] Al-Ash'ari also reports that he rejected finite divisibility and said, "there is no part without parts and no half that cannot be halved and that every part can be subdivided indefinitely."[83] Al-Khayyāt says that, "Ibrahim [al-Naẓẓām] denied that bodies are composed of indivisible parts, and

78 *Maqalāt*, 2/235.
79 Ibid. 160.
80 Ibid.
81 See *Kitab al-Haywan* and *Rasa'il al-Jaḥiẓ*.
82 Al-Ash'ari, *Maqālāt*, p. 318.
83 Ibid.

he claimed that there is no part of the bodies that cannot be envisioned to be divided into two [parts]."[84]

The important thing here is how al-Naẓẓām agreed that bodies are finite in the totality of volume and area but assumed that they can be envisioned as being subdivided indefinitely. The question now is: has al-Naẓẓām denied the existence of the indivisible part ontologically or was it that he believed (just because one can always imagine any part, no matter how small) it to be subdivided into smaller parts? This question can only be answered in conjunction with his theory of motion, which I will shortly discuss. But let us first ask: what was the motivation for al-Naẓẓām's negation of the finite divisibility of bodies in the first place? This, I will try to answer once we understand his views concerning transients.

AL-NAẒẒĀM ON TRANSIENTS

It is reported that al-Naẓẓām negated the existence of transients (*aʿrāḍ*) and was inclined to assume that it is the nature of beings which drives their interactions with other beings in the world. This, perhaps, was elaborated in his theory of causality, which is based on the notion of potentiality (*kumūn*). Al-Naẓẓām was thought to consider transients as bodies, and the only transient he acknowledged was motion. Al-Ashʿari says, "al-Naẓẓām said that it is impossible for the transients to be realised and that there is no transient other than motion."[85] And it seems that al-Naẓẓām has affirmed that transients cannot be found to stand alone.

Also, al-Ashʿari reported al-Naẓẓām as saying, "the *aʿrāḍ* do not oppose each other; what does so are bodies like heat and cold, black and white, sweet and sour, and these all are corrupting bodies that corrupt each other, and so it is the case with any two corrupting bodies, that they will be contradictory."[86] The statement above is not so clear as to give us the intended meaning. However, I understand it to mean that al-Naẓẓām is attributing the properties of the bodies to the bodies themselves, not to the mutable transients. As I said, this is because al-Naẓẓām seems to have adopted the belief that it is the nature of the body which decides how it will interact with other objects in the world. It also appears that he did not want to explain this by going into the microscopic level, as he would then face the question of atomism and finite divisibility, which he did not accept.

In fact, the position of al-Naẓẓām on atomism is reflected in his understanding of motion, as seen by his opponents arguing against his denial of

84 Al-Khayyat, *Intisar*, 76.

85 Maqālāt, 362.

86 Maqālāt, 376.

atomism by challenging him to explain motion. They raised Zeno's paradox against him, arguing that if the trajectory of motion is to be a continuous line, then moving bodies would not be able to cover the distance between any two points; that is, if distance is assumed to be composed of an infinite number of parts. This is because, in order to cover any distance, the body has to cover half of it first and then cover half of the remaining distance again, and then half of that again, and so it goes on *ad infinitum*. This means, essentially, that the moving body will never be able to cover any distance at all.

Al-Naẓẓām is reported to have solved this paradox by claiming that the body moves in leaps or jumps (*ṭafra*), so that it would not pass through all the points along the trajectory of motion but jumps from one place to another without passing through the distance in between. Most of the Mutakallimūn rejected this notion of *ṭafra* and considered it absurd. Ibn Ḥazm, who rejected atomism, also rebuffed the notion of *ṭafra,* considering it something impossible. I will come back to this subject in Part III of the book, when I will discuss the concept of motion in *kalām*. Unfortunately, the acclaimed book of al-Naẓẓām is lost, otherwise we could have understood this argument about atomism more clearly. In any case, I will also return to discuss this problem in Part III of this book, when presenting views about motion in *kalām*.

IBN ḤAZM AND ATOMISM

Ibn Ḥazm al-Ẓāhirī was one of the great theologians of the 11th century. His influence in Islam has extended over many centuries and, on the whole, he is known to have an independent approach to the deduction of Sharīʿa laws. Ibn Ḥazm adopted some of the *kalām* ideas, arguments and principles but opposed the notion of *jawhar* (not discreteness), despite believing that the properties of objects are emergent transients. He also believed in the relative nature of space and time, and rejected the notion of absolute space and absolute time suggested by Greek philosophers. Nevertheless, he criticised the early leaders of the Muʿtazilis, including al-Naẓẓām. He also spoke out against Abū al-Hudhayl and the Ashʿaris for their claim regarding the renewal of transients, particularly the spirit.[87]

As for atomism, Ibn Ḥazm also devoted many pages to criticising this concept—particularly the notion of a singular *jawhar*—discussing the existence of the *jawhar*, and presenting his arguments against it. First of all, Ibn Ḥazm refuted what Abū al-Hudhayl said about the power of Allah to disintegrate the body until it became indivisible parts. This statement, he found, limited the power of Allah by ending his power to disintegrate at the level of the *jawhar*.

87 Fisal, 4/58.

Theologically, God is omnipotent, and his power has no limits. Aside from this, it seems that Ibn Ḥazm was not convinced with the ontological status of the *jawhar*, as we will see in his arguments.

He attacked the notion of the singular *jawhar* with five arguments. The first says that negating the *jawhar* does not necessarily mean bodies will have an infinite magnitude. In this respect, he counters the claim of the Muʿtazilis that the *jawhar* has no magnitude, saying that this would imply that the bodies, which are composed of such entities, would certainly have no total magnitude. But here, Ibn Ḥazm is ignoring the fact that those who claim the *jawhar* has no magnitude are doing so because they define magnitude as being an intrinsic property of the *body* specifically. However, since the *jawhar* is not a body, his objection does not apply. On the other hand, it is known that those who denied the magnitude of *jawhar* agreed that, once it combined with another, it gained both magnitude and length, and that that is how bodies of finite extensions are formed. Al-Ashʿari reports, "If the indivisible part is combined to another then each of them is a body as long as they are in combination, if they get separated, none of them is a body."[88] He attributes this statement to ʿIssa al-Sūfī of the Baghdādī Muʿtazilis. It seems that Ibn Ḥazm has either overlooked this opinion, or, most probably, ignored it.

This description means that magnitude arises as a result of the integration of the parts which themselves are without magnitude. This is certainly an advanced vision and could be considered a forerunner of the notion of differential calculus. The differential increment has no magnitude on its own but would acquire a magnitude on integrating with other parts. Physically, this can also be understood to mean that the magnitude of the *jawhar* is obtained upon mutual interaction with other jawāhir. The great value and implication of these thoughts have only begun to be realised in this modern age of science.

The second objection of Ibn Ḥazm is to say that for the body to have a finite measure, there should be a non-divisible part at which the finite measure stops. He says that this is a corrupt illusion and that, "We do not say that the finiteness is caused by the measure, but we say that for every body there is a boundary and a surface at which its extensions end, and that end, at which the measure stops, is a finite part, but it can be subdivided further."[89] This argument is not strong enough to refute the claim of there being a non-divisible part because Ibn Ḥazm admitted that the finite measure of a body implies that it has a finite number of units. The last of these units is where the measure

88 *Maqalāt*, 2/235.

89 Fisal, 59.

ends, but then he claims that this last part could, potentially, be divided further. This notion becomes clearer in his third argument, where he admits that infinite divisibility is *practically* impossible despite it being in the power of God to perform infinite disintegration. He says, "we do not object [to the fact] that the fine parts of the flour no creature in the world can divide it."[90] So here, Ibn Ḥazm admits that it is *practically* impossible to divide very fine parts of a substance any further. This implies that he would accept having a certain minimum but cannot find a reason for this minimum to be envisaged as being indivisible.

In the fourth argument, Ibn Ḥazm presents his objection to accepting the concept of *jawhar*, saying, "it is not true to claim that we say that whenever a body covers a place by moving, it means that it has covered an infinite extension."[91] Then he says, "we say that every part that is an actuality should be finite in number, no doubt. We have not said that its finitely divisible parts are existing; this is untrue and impossible."[92] These statements clarify that Ibn Ḥazm refutes the reality of an infinite, as well as denying that the infinite number parts, which it can be imagined are produced by dividing a finite body, can really exist. Instead, he stresses that such a divisibility can be envisaged only. He further asserts that, "But once actually produced, either from things or in time, by division into parts, then all this should be finite in number… and whatever has not been actually produced cannot be considered something or a number of numerable."[93] Here, we see clearly Ibn Ḥazm negating the actual possibility of infinite divisibility, while affirming the possibility of it occurring theoretically. Quite clearly this man is an empiricist. This kind of understanding was an advanced step in comprehending the differences between theory and practice. Should it have been adopted a long time ago, many problems in understanding the world would have been solved much earlier.

Ibn Ḥazm confirmed his empiricism when he disvalued the question, "which of the two has more parts: the mountain or a mustard seed?" by answering, "unless the mountain is divided, and the seed is not divided, then we have no parts." He used this in his fifth argument where he responded to the question put forward by some Mutakallimūn on the knowledge of God of the number of parts in a mountain and in the mustard seed. He says that this is, again, a silly question because, "God does not know the number of hairs in the beard of a bald person, and He doesn't know the number of children of an

90 Fisal, 60.
91 Ibid. 5/60.
92 Ibid. 5/61
93 Ibid.

impotent."[94] In this argument, Ibn Ḥazm is stating the rule that whatever has not been actualised in the world, or deemed impossible to happen, cannot be taken into account in argumentation.

In summary, I can say that Ibn Ḥazm considered the body, as a whole, to stand for the *jawhar*, representing its essence ontologically, while accepting the notion of transients and associating them with the body itself. His refutation of the existence of the indivisible part was based on the ontological impossibility, not the epistemological contingency. From what we gather about his approach, Ibn Ḥazm was an ideal empiricist. This was indeed his known character that he affirmed with his philosophical analysis. This trend can also be found in his theological studies as a cleric, where he considered the direct meaning of the texts and their apparent intentions rather than any implicit meanings. For this reason, he is known among theologians to be concerned with the apparent, or manifest, meanings of Qur'anic verses (*Ẓāhirī*).

Ibn Rushd on Atomism

Ibn Rushd (Averroes) was a great rational thinker, who adopted the methodology and views of Greek philosophers, mainly Aristotle. He translated and explained a large volume of Aristotle's works, especially his metaphysics. Ibn Rushd criticised *kalām* atomism from an Aristotelian point of view, asserting the infinite divisibility of things. His main arguments were:

1. That the existence of an indivisible part is not known by itself.
2. That the concept is equivocal and there are contradictory views about it.
3. That it is not for *kalām* methodology to deal with such a topic, but it is for the demonstrative methodology to provide analysis and proofs for such a problem.
4. That the Ash'aris, in particular, have mostly used primitive discourse to prove atomism and these have shown no demonstration of proof.

Ibn Rushd says, "The first premise, which says that the jawāhir cannot be devoid of their transients, if they mean by it the self-dependent bodies, then it is correct, but if they mean by the *jawhar* the indivisible part, which they call the singular *jawhar*, then there is much doubt about it because the existence of non-divisible *jawhar* is not self-evident and there are many contradicting opinions about its existence, and it is not in the capacity of *kalām* discourse to deduce the truth out of those views, but that is for demonstration (*burhan*), and those who master this are very few."[95] Clearly, Ibn Rushd was not satisfied with the proofs provided to demonstrate the existence of the singular *jawhar*.

94 Ibid. 62

95 Ibn Rushd, *Adillah*, 35.

The reason for this, it seems, is that he only acknowledges Aristotelian methodology in demonstration.

Ibn Rushd also discussed the arguments of the Ashʿaris which they presented in support of their views, "Their famous demonstration is they say: that from basic information we know that an elephant, for example, is larger than an ant because it has more parts than the ant. If so, then the elephant is composed of those parts and is not one whole unit, and if disintegrated, the body will be those parts and if composed, it would be out of those parts."[96]

To counteract this argument of the Ashʿaris, Ibn Rushd said, "But this mistake came through their analogy with a discrete quantity and applies to the discrete, which one should equally apply to the continuous."[97] Then he continued, "But this could be true for numbers. I mean one number is more than another because of the differences in their units, but as for the continuous whole, this does not apply. This is why we say in respect of the continuous whole that it is greater or larger and we do not say that it is more or less, but we say for the number that it is more or less, and we don't say it is larger or smaller."[98] The main point of this statement is that Ibn Rushd considered objects to be one whole and assumes they cannot be envisaged as being composed of smaller units. The argument behind this statement is that numbers are discrete, whereas geometry—describing, for example, the shape of a whole body—is continuous. This is why Ibn Rushd concluded that if the assumption of discrete composition of things were adopted, then, "all objects will become numbers and there can be no continuous whole at all. Consequently, the science of geometry will become numerology."[99]

About the Conclusion of Ibn Rushd

The above conclusion of Ibn Rushd is one of his most elegant deductions and reflects his genius. Although this does not disprove the argument of the Ashʿaris on the singular *jawhar* and atomism, the conclusion of Ibn Rushd, which says that if discreteness is adopted then the geometrical presentation of structure will turn into numerical presentation, is quite true. This, indeed, was demonstrated when atomism and discreteness were introduced at the beginning of the 20th century.

When Niels Bohr introduced the principle of quantisation (discreteness) to the orbital angular momentum of the electron in his model for the atom, it enabled the representation of the orbits of the electrons, which were normally given in geometrical terms—say a circular or elliptical orbit—to be described

96 Ibid.
97 Ibid. 35.
98 Ibid.
99 Ibid. 37

by numbers: the so-called 'Quantum Numbers.' So, instead of describing the orbit by saying that it is a circular orbit, for example, we need only call it an l=0 orbit or an s orbital. And instead of defining another orbit as elliptical in shape, it is usual to say it has l=1, and so on. Clearly, in this presentation, geometry is described by numbers and, consequently, becomes a question of numerology. Most of the calculations in atomic physics take quantum numbers as prime indicators of structure as well as the values of the physical quantities concerned.

This valuation of Ibn Rushd, from the viewpoint of modern scientific understanding, is by no means a projection of our insight into an old-age statement. It is, instead, an intellectual achievement that could not be evaluated at the time it was made. Should Ibn Rushd have ended his statement saying that there could be no continuous whole period, then I would not be able to take his statement this far but since he concluded that discreteness implies the conversion of geometry into numbers, we find that it certainly resonates with our modern times. However, it is unfortunate that Ibn Rushd refused to accept that geometry could become numerology, and thus denied discreteness. Besides this, we should also note that the validation of Ibn Rushd's conjecture gave support to the old atomism of the Mutakallimūn, in one way or another, within the context of the argument presented.

SHLOMO PINES ON KALĀM ATOMISM

Shlomo Pines wrote a book about Islamic atomism in which he analysed and discussed the premise of Islamic atomism, as far as he could find it in the original writings of the Mutakallimūn. One of the aims of the book was to investigate the origins of Islamic atomism, which some claim have Greek and Indian origins. Pines investigated those theories in depth and found insufficient evidence to accept this claim (except for some similarities which he found in a number of viewpoints).

I have noticed that one of the motivations for Western scholars studying sources of *kalām* atomism was their puzzlement at the genius of early Muslims. With regards to this, Pines says, "the suggestion that this theory [*kalām* atomism] arose quite independently within Islam must presuppose an intense theoretical interest in philosophy and science on the part of the earliest Mutakallimūn."[100] Clearly, from the start, Pines had the preconception that Muslims were not qualified to pursue their atomic theory independently. This might seem logical for a nation that had no philosophical legacy. But we should remember that this theory in *kalām* was composed from the 2nd to 8th

100 Pines, Atomism, 108.

centuries, during which time Muslims had established a large empire extending from China to Spain. Nevertheless, Pines continues his analysis, saying, "the theological import of the concept 'atom' cannot by itself explain the very complicated form in which some particular problems, that have nothing to do with theology, arise early on. The problem of the minimum number of atoms required to make up a body may serve as an example. The complexity itself indicates that the issues had been developing over a long period of time." It may be correct to say that the word 'atom' in the Qur'an has no intellectual implication whatsoever, but we should not ignore other facts related to the social and intellectual transformation that Islam has produced. In respect of this, let us continue with what Pines wrote, that, "On the basis of extant accounts, one can hardly speak of an initial theoretical curiosity of such intensity among the earliest Mutakallimūn."[101] Indeed, the basic terms 'atom' and 'atomisation' might have been imported from elsewhere but surely the development of this concept, which was highly sophisticated, makes it clear that Muslims had completely departed from the Greek model, as set by Democritus and Epicurus.

Muslims may have been in touch with some of the works of Greeks and Hindu scholars, and, no doubt, some ideas might have been picked up from here and there, but we should not ignore the fact that the teachings of the Qur'an equipped Muslims with a very rich and solid background in respect of the worldview. Muslim scholars, who would wake up before sunrise, wash their bodies and perform their religious rituals so early in the day, had much time to spare for studies and learning, which usually started before sunrise. Up until noon, a Muslim might achieve a great deal of work during these six hours. Working after lunch for two or three hours, a Muslim would have done more than eight hours of work. This is part of the greatness of the lifestyle of a true Muslim. So, it should be no surprise that Muslims grasped so much knowledge and possessed an intense, theoretical curiosity, such as we see among the earliest Mutakallimūn. Those who are curious about the achievements of Muslims during the early centuries of Islam should learn about their system of work and study during that time. Besides this, as I mentioned above, the Qur'an is not a book of stories or legends. It is a book of knowledge containing signs which act as a distinct and accurate guide for those who think, and for people of knowledge. This argument is strongly supported by the fact that the Mutakallimūn always made the truth of Islam their starting point and, consequently, had to begin with the intellectual construct provided to them by the Qur'an; this is more than clear in the works of the Mutakallimūn.

The astonishment of Pines about that intense, theoretical curiosity of early Muslims and their involvement with complicated philosophical topics

101 Ibid.

is repeated in the views of another esteemed scholar, Harry Wolfson, whom I will present in the next section. However, Pines was almost impartial in his study when he exposed the differences between the Greek atom and the Islamic atom, and concluded, "There may be a certain resemblance between the minima (not the Atoms) of Epicurus and the *ajzā'* of the Mutakallimūn, especially as the *ajzā'* are contrasted with bodies; and atoms also count as bodies. One could also establish a certain similarity between the theories of time and space in the two systems. However, the scantiness and uncertainty of our knowledge of that side of Epicurus theory do not allow us to press these analogies any further with the hope of arriving at any sound conclusions."[102] Pines also says, "O. Pretzel adopts the thesis that the atomism of *kalām* came to be by means of a thorough transformation of the Greek theory… but as it seems to me, Pretzel's argument in support of his conjecture is not sound."[103] After a lengthy analysis of Indic sources, in search of the origins of *kalām* atomism, Pines concludes that, "the only answer that we may give at this time to our problem, namely Indian influence upon Islamic atomism, is that such influence is not outside the bounds of possibility; but the state of our knowledge at present does not allow us to give a positive answer. Our inquiry thus ends with a *non liquet* (it is not clear)."[104]

HARRY WOLFSON ON ISLAMIC ATOMISM

Harry Wolfson is a Harvard scholar who studied *Kalām* extensively and published a magnificent book entitled, *The Philosophy of the Kalām*. In this book, he devoted about eighty pages to discussing *kalām* atomism. He held the view that Islamic atomism originated with the Greeks. His starting argument resorted to pointing out the similarity between words, where he sometimes tried to establish unfair, and even inaccurate, relationships between words that have different meanings in Arabic and Greek, such as the word '*jawhar*', which he thinks corresponds to the word 'essence.' In fact, this is a bad example since *jawhar* means 'essence', and not substance. It seems that Wolfson, and other such scholars, have done this under the influence of their pre-conceptions in order to prepare the reader to accept their conjectures, even if not enough verification is provided.

Besides this, some questions, which have been taken out of context, give the impression that a number of Muslim scholars referenced non-Islamic sources of atomism. For example, on pages 466–467, Wolfson mentions Ibn Ḥazm as having testified that, "the atomism of *kalām* is to be traced to some

102 Ibid. 113
103 Ibid. 115
104 Ibid. 140.

of the ancients." In fact, this statement was taken out of context as Ibn Ḥazm actually says, "Al-Naẓẓām and everyone who experienced the thought of the ancients believed that there is no part, even it is small, that can be subdivided indefinitely." By no means should this statement be understood as a testimony by Ibn Ḥazm that atomism of *kalām* can be traced to the ancient Greeks.

In any case, we find Wolfson expressing astonishment that Muslims accepted atomism while most Greek philosophers rejected it. However, he also exposed fundamental differences between Islamic and Greek atomism, such as temporality and finiteness in number, and that the Islamic atom lacks magnitude while Greek atoms, which have magnitude, are eternal and of an infinite number. Despite this, Wolfson, as with many other authors, failed to recognise that the Islamic atom is an abstract entity that actualises on combining with other atoms, while the Greek atom is a real and stand-alone object.

Islamic atomism is but one principle out of several that forms a consistent system of thought, as shown in the chapters of this book, whereas the Greek atomism is one, stand-alone theory of matter. This makes the two theories fundamentally different. Greek atomism is a view that has nothing to do with theological beliefs and, therefore, is a description of the ontological construct of matter, whereas Islamic atomism is a much deeper epistemological concept that is strongly related to religious belief and the worldview of Islam.

However, it is surprising that after he found a negative result for a Greek or Indian source for the Islamic atom, Wolfson resorted to his imagination in order to explain the sudden intellectual development of the Arab-Islamic mind, and its intense, theoretical interest in complicated issues, such as atomism and causality. In doing so, he returned to a Greek origin, looking for sources of knowledge in Harran (north of Syria), where a number of intellectual schools were at work at the dawn of Islam: around the 6th and 7th century. Wolfson looked for bits and pieces of reports and doxographies, even distorting the meaning of words sometimes, to prove his conjecture of the Greek origin of *kalām* atomism.

In any case, I feel that this methodology of "conjecture and verification", as defined by Wolfson, does not apply when studying such topics since, in these subjects, we always have many grey areas. Instead, I find it is more efficient to use the approach that looks at all the other factors contributing to the question we are dealing with. We should examine the social and intellectual change that took place in the life of Arabs after Islam, then we will learn where that burst of intellectual richness came from.

AL-NOOR DHANANI ON ATOMISM

Al-Noor Dhanani studied *kalām* atomism as part of his investigation into *kalām* and Hellenistic cosmology for his Ph.D. under the supervision of A.I.

Sabra at Harvard. Dhanani's study was published in a book entitled: *The Physical Theory of Kalām*. The book is a detailed piece of research into several fundamental concepts of atomism, space and voids in Bastian Muʿtazilis cosmology. Dhanani used an excellent collection of original sources for the presentation of thoughts, but his use of certain terms sometimes distorted their actual meaning. For example, he used the term *atom* to mean the term *jawhar*. This could cause confusion since the *jawhar* is not meant to be the atom. Instead, it is the indivisible part, while the atom should be identified as the *jawhar plus* the transients. Ignoring the fact that the *kalām* atom is composed of an indivisible part, which is the *jawhar*, and the accompanying transients makes it difficult to understand the basic difference between the Greek atom and the *kalām* atom.

In addition to this, assuming that the *jawhar* has magnitude and ignoring the assertion of many *kalām* scholars that the *jawhar* has no magnitude, as well as contemporary studies such as those of Pines and Wolfson, assertions that the question of magnitude is a vital difference between the *kalām* atom and the Greek atom, the study of Dhanani appears much distorted and confusing. Moreover, claiming that the *jawhar* occupies space is a fundamental error in understanding the term *taḥayyuz*, which the Mutakallimūn used to explain the being of the *jawhar* and its ontological meaning as a basic constituent of the body. However, using the term *accident* to describe the term *ʿaraḍ* might be acceptable, but to use the word *substance* to translate the term *jawhar* is something that leads to much confusion, especially when we encounter the term *al-jawhar al-fard*. The Arabic meaning of the word *jawhar* is 'essence', but this word is not quite expressive of the concept of *jawhar*, which is why I have used the Arabic term in this book.

The other problem with Dhanani's work is his misunderstanding of some of the original Arabic texts that he studied. One of these is his misrepresentation of al-Ghazālī's attitude towards *kalām* atomism.[105] Al-Dhanani understood that al-Ghazālī probably rejected the discrete cellular space of *kalām* atomism. This, he concluded on reading al-Ghazālī's *Tahāfut al-falāsifah* and *Maqāṣid al-falāsifah,* in which he says, "there is nothing in the premises of geometry and arithmetic which is contrary to reason." By no means can one take this sentence, which has been taken out of context, to mean that al-Ghazālī rejected *kalām* atomism. When we go back to the original text, we see that al-Ghazālī was actually listing the problems he saw in the contradictions of the philosophers:

> These, then, among their metaphysical and physical sciences, are the things in which we wish to mention their contradictions. Regarding mathematical

105 Al-Dhanani, *Impact.*

> sciences, there is no sense in denying them or disagreeing with them. For these reduce in the final analysis to arithmetic and geometry. As regards the logical [sciences], these are concerned with examining the instrument of thought in intelligible things. There is no significant disagreement encountered in these.[106]

Finally, al-Dhanani concludes that al-Ghazālī had a lukewarm commitment to *kalām* atomism and that his response to Ibn Sīnā's critique of *kalām* atomism was to abandon it. This conclusion is erroneous, as we find that al-Dhanani was inaccurate when quoting and interpreting al-Ghazālī.

ATOMISM AND TEMPORALITY

Is atomism an independent principle of *Daqīq al-kalām* or is it related to the other principles? Ibn Matawayh asked the same question in his book, *al-Tadhkira*, "Does saying that the part can be subdivided breach the claim of the temporality of the body so that those who believe in its [infinite] divisibility would acknowledge its temporality?."[107] Ibn Matawayh then answered his own question, saying, "the claims that the temporality of the body was based upon does not need proof of the indivisible part… But whoever believes in the infinite divisibility of a body is obliged to believe in the eternity of the world since an infinite cannot exist, for how can such an infinite be temporal when it was once believed to be unlimited? This might compel one to believe in its eternity." Clearly, Ibn Matawayh was very particular in composing this answer. He wanted to convey that the difference between the dependence of the arguments is not the same as the dependence of the results, for the argument of temporality is independent of the argument of atomism. But not believing in atomism compels one to believe in eternity and deny the temporality of the world. However, the argument of Abū Al-Hudhayl about atomism when he says, "God can disintegrate a body to make its parts non-divisible", sounds as though there is a relation between temporality and atomism. Atoms, according to this argument, have to be temporal, otherwise it cannot be understood how they come into being.

On the other side of the question, we know that *kalām* atomism has associated transients with the *jawhar*, and since most transients do not endure, they therefore have to be temporal, and since the *jawhar* do not exist without transients, one can establish that there is a connection between atomism and temporality.

106 Al-Ghazālī, *Tahāfut*, 11.
107 *Tadhkirah*, 171–172

THE ATOMISM OF TIME

The atomism of *kalām* covers everything, including time and motion. This poses a very important problem in contemporary natural philosophy, namely modern physics. The Mutakallimūn envisaged time as being discrete, with the finite duration being divided into minute, indivisible parts, each of which is called an *an*. An *an* represents a stationary moment in the present.

Perhaps the best description for the discreteness of time was given by Ibn Ḥazm when he said, "and the finiteness of time exists by what follows after that which has passed, and the annihilation of any moment of time after it has existed and the beginning of what comes afterwards. This is because every moment of time ends now, as this is the border between two times, at which the past ends and the future starts, and so every duration of time is composed of finite parts which begins as we said before."[108] Ibn Ḥazm had a clear definition of time for which he had a known reference. This was natural astronomical time. He said of it, "and in our conventional time is the duration of the existence of an object being at rest or in motion, or the duration of the existence of a transient in the body. Generally, we can say it is the duration of the orbit and whatever is related with it." However, according to Jurjānī, the Mutakallimūn defined time as "a known renewable measuring another envisaged renewable, as they say: I will come at sunrise; that the sunrise is known and his coming is envisaged, and once the envisaged is combined with the known [meaning the occurring event] envisagement is nullified."[109]

The discreteness of time is embodied in the Arabic language by the timing of an action. Al-Zajaji said, "the verb is of two kinds: a past and a future. The future has not happened yet and has not been endowed with elapsed time nor has it been yet, and the past is what has been covered with at least two instances of time; one instant at which it was and the next where it has been told about. The verb of present is the one which is composed at the time of the speech itself."[110] Perhaps this explains why the Mutakallimūn differed on the understanding of time but agreed on it being atomised. This statement of Zajaji shows that time atomism was a structural part of Arabic; a very important note since language is known to be the paradigm of thought.

The contemporary significance of the atomism of time appears as a necessity towards the integration of physical theories. As we now know, Quantum Theory offers the best description of the microscopic world. Through it we have been able to achieve great advances in technology using the microscopic description of matter. Laser techniques, with their wide spectrum of

108 Fisal, 1/57.

109 Tarīfāt: al-Zamān.

110 Al-Zajaji, Iḍaḥ, 86–87.

applications in industry and medicine, are the fruitful result of laser physics, which is based on understanding quantum transition of different types which occur in atoms and molecules.

On the other hand, the Theory of Relativity has enabled us to view the microscopic world in a new way, one in which space and time are fused together to form the spacetime continuum. The relationship between mass and energy was discovered by Albert Einstein himself, and it was realised soon afterwards that this connection could explain the vast energy stored in microscopically-bound systems, such as atomic nuclei. This discovery enabled the world to utilise nuclear energy and construct large power stations with virtually unlimited resources of energy. Thus, the Theory of General Relativity has empowered us with a better and more accurate understanding of the universe.

The unification of quantum mechanics and the Theory of General Relativity has become inevitable in order to achieve the next major step in physics and the understanding of nature on both the microscopic and macroscopic levels. Such an achievement will enable mankind to discover new technologies; ones which make all the advanced technologies of today part of the past. In doing so, the quantisation of time becomes a necessity to enable the unification of quantum mechanics and the Theory of Relativity.

KALĀM ATOMISM AND THE MATHEMATICAL CONCEPT

Whenever we come across an evaluation of the works of the ancients, we should consider the topic in its context and avoid projecting our present knowledge on explaining it. However, this should not mean that we underestimate or devaluate contributions from the past, using the excuse that those people did not have an advanced understanding such as ours. No, this is not always true, since, at times, we come across certain topics which have preserved their full meaning down through the ages. Take, for example, geometry and calculus. Although calculus is a relatively recent invention of the 17th century, geometry is a technique that has been preserved for several millennia. The connection between geometry and calculus enables one to obtain results, which were once thought to be obtainable through calculus only, using geometrical techniques. This relationship enabled the Babylonians in the 4th century BC to calculate the position of the first heliacal rise of the planet Jupiter using a geometrical approach. They achieved this by calculating the area under the curve which was obtained by plotting the velocity of the planet versus its position in the sky. This magnificent technique is considered to be of a highly scientific value in our modern times.[111] It would be silly to say, for example, that the Babylonians

111 Mathieu Ossendrijver, Ancient Babylonian astronomers calculated Jupiter's position from the area under a time-velocity graph, *Science*, Vol. 351 (6272), January 2016.

did not mean to calculate the area under the curve just because they did not know of Cartesian coordinates.

Kalām atomism assumes a sort of atom that has no magnitude, describing it as existing because of its occupancy (*taḥayyuz*) rather than its occupying space like any other object. Furthermore, it maintains that such atoms do not acquire volume or magnitude unless combined with another atom. This compels us to describe this atom of *kalām* as an abstract object, unlike the atom of Democritus or Epicurus. The concept of time atomism, which was suggested by the Mutakallimūn, is intimately connected to the understanding of motion and the integrity of space and time, for these concepts are inter-related according to the physical theory of *kalām*. Perhaps it is of some importance to point to the discussion presented by al-Ijī in his book, *Mawāqif*, but in any case, I will go back to discuss these topics in Part III of this book.

ATOMISM OF DEMOCRITUS AND EPICURUS

The atoms described by Democritus are tiny, solid balls that are indivisible, infinite in number, eternal, and do not carry transients necessarily. Because of their different shapes and the way they form a composition, these atoms make up the different kinds of matter. The atoms of *kalām* are abstract entities, almost mathematical forms of the lowest level of existence; they are all the same, apart from the transients they carry. The transients are what make the properties of whatever is composed of these atoms different.

Epicurus maintained that atoms were indissoluble and eternal, but that they were not the ultimate components of matter. Atoms, according to Epicurus, are extended bodies that have a certain size and internal parts. These parts (called minima) have no further parts, nor can they exist in isolation. Instead, they subsist eternally within the atoms whose parts they are. It is also reported that Epicurus believed in the discreteness of space, time and motion. However, the most important texts for the elaboration of this theory appear to be lost.[112]

A number of authors have tried to establish an analogy between the minima of Epicurus and the *kalām* atom. Dhanani, for example, has meddled with this analogy extensively,[113] but Pines was much more conservative on this issue, saying, "there may be a certain resemblance between the minima (not the atoms!) of Epicurus and the *ajzaʾ* of the Mutakallimūn, especially as the *ajzaʾ* are contrasted with bodies, and atoms also count as bodies. One could also establish a certain similarity between the theories of time and space in the two systems. However, the scantiness and uncertainty of our knowledge of that side of Epicurus' theory do not allow us to press these analogies any further with

112 Pines, *Atomism*, 112.

113 Dhanani, *The Physical Theory of Kalām: Atoms, Space, and Void in Basrian Muʿtazilī Cosmology.*

the hope of arriving at any sound conclusions."[114] I would say that these analogies between Epicurus' minima and the *kalām ajzā'* have sometimes been taken too far in an endeavour to demonstrate the Greek origin of *kalām* atomism. This is a far-reaching goal that cannot be achieved unless we take the legacy of *kalām* out of its social, religious and intellectual context.

OUR CURRENT UNDERSTANDING OF ATOMISM

The question of discreteness and continuity in the structure of the world is a persistent one which has been discussed and analysed through the ages. In the past, it was part of natural philosophy. Nowadays, it is at the core of science, despite the fact that science can only deal with empirical facts. In any case, whereas classical physics (the study of physics pre-20th century) established the paradigm of continuity, modern-day physics (of the 20th century) established a new paradigm based on discreteness. Thermal radiation energy was the first physical quantity to be atomised when data gathered during laboratory experiments obliged physicists to consider a new vision in order to understand the behaviour of thermal radiation.

These observations forced the German physicist, Max Planck, to conclude that radiation contained within a spherical cavity, with a small opening, behaved like discrete corpuscles instead of continuous waves, as had been thought for a long time. Soon after this discovery, another German physicist, Albert Einstein, found that such a vision could be utilised to explain another phenomenon; namely, the photo-electric effect, by which electrons are known to be ejected from a metal surface when an intense, energetic light beam is projected onto it. Accordingly, Einstein proposed that light, like thermal radiation, is made up of energy quanta, with the amount of energy in each quantum being proportional to the frequency of the radiation.

It was John Dalton (1766–1844), the English chemist, who revived the concept of atomism in understanding the structure of matter. He pioneered the development of modern atomic theory. Although modern atomic theory has become much more involved than Dalton's original theory, its essence remains valid.

An interest in studying the internal structure of atoms, in light of the discovery of negatively charged particles (electrons) as constituents of matter, encouraged Lord Rutherford to probe the atomic structure of a piece of thin gold foil using positively charged particles (α- radiation), which were emitted from some radioactive material. This famous experiment demonstrated that atoms have a central, internal structure, called the nucleus, in which the

114 Pines, *Atomism*, 113

positive particles (protons) are located within a tiny place, with the negatively charged particles (electrons) floating around outside.

However, as experiments had already shown that positive and negative charges attract each other, it was a problem to solve how the negatively charged particles can float outside the positive nucleus. It was natural to assume that the negative electrons revolve around the nucleus in closed orbits. However, the electromagnetic theory of James Clark Maxwell suggested that these electrons would soon fall into the nucleus once they had radiated their energy (electric potential energy), which they had acquired by being in their positions. This marked a failure for the new atomic structure.

Two years later, the Danish physicist, Niels Bohr, found a solution to this dilemma. He proposed that certain orbits 'enjoyed privileges', which allowed the electrons in them to rotate without radiating energy; that is, they behaved as if they were locked into those orbits so that no radiation was emitted. Once the electron moved out of its orbit into another, it radiated a quantity of energy that was equal to the difference between the energy levels of the electron in the two different orbits. This move from one orbit to another (called transition) had to be done in jumps rather than in a continuous manner. The Bohr model of the atom was based on assuming that, along with their energy, the angular momentum of electrons in the atom was also discrete. Bohr was then able to explain the spectral series obtained on heating chemical elements and compounds. The quantum jump, however, presented a dilemma for some scientists, including Erwin Schrödinger.

These discoveries, being successful in efficiently explaining several experiments and phenomena, established the so-called quantisation of physical quantities, such as energy and angular momentum. But soon physicists realised the need for a comprehensive theoretical framework that could explain everything within this new vision. Why should there be quantisation? What was the physical foundation of the particle-like behaviour of energy, which was always described as waves and wave motion? These questions became much more serious when electrons were found to exhibit wave-like properties. In a famous experiment called 'electron diffraction', the puzzle became double-sided; displaying a sort of duality which involved particles and waves exchanging roles.

The puzzle proved difficult to solve until French physicist, Louis de Broglie, expressed this duality using a simple relationship, which described wave properties in terms of particle properties. He suggested that a particle, moving with a given momentum, should have an associated wave of some kind, the wavelength of which was inversely proportional to the momentum of the particle. This correlation helped to push the mathematical formulation of a comprehensive theory of quantum behaviour a major step forward. This step was taken by Erwin Schrödinger, who discovered the equation of motion for a dual entity, and which was called "the wave-packet." The Schrödinger equation described the time development of a quantum state that enjoyed the dual properties of both particle and wave.

Meanwhile, there was an important development at the conceptual level where the 'wave-packet' concept was used to describe the new dual entity. Particles were envisaged as a superposition of a large number of waves combined to construct a 'group wave' enveloping those superimposing waves, which were called 'phase waves.' The ideal group wave, representing a particle in a given locality, is represented by a bell-shaped pulse. The more phase waves contribute to the group, the less wide will be the bell. It turned out that the width of the bell is directly related to the uncertainty in the position or momentum of the particle. A point particle is represented by a vertical line perpendicular to its position in the coordinate space. This is a pulse which is composed of an infinite number of phase waves.

It was also known that this representation of a particle in the coordinate space could also be visualised in another space called the 'momentum space.' This is achieved by using a mathematical technique called 'Fourier Transformation.' Through this, a wave packet's width in the momentum space is related to its width in the coordinate space, where the product of the first and the second is constant. This means that if the width of the wave packet in the coordinate space increases then its width in the momentum space decreases, and vice versa. Using the de Broglie relation, it was an easy task to determine that the width of the wave packet in the coordinate space represents the minimum error when measuring its position, whilst its width in the momentum space represents the minimum error in its momentum. This established the rule that the product of the uncertainty, when measuring the position and momentum of a particle simultaneously, is equal to Planck's constant, a minute quantity that plays an essential role in quantising energy and angular momentum, as shown by the laws of Planck and Bohr. This correlation between the natural error in measuring the position and momentum of a particle established the Heisenberg Uncertainty Principle; a restriction that played a very important role in the structure of the universe and uncovering the indeterministic character of the natural world.

The Schrödinger equation was devised to describe particles moving with low velocities when compared with the velocity of light (called non-relativistic particles). It was realised that we needed a relativistic version of the equation of motion of a quantum particle. This equation was discovered by Paul Dirac and was the fruitful result of introducing the Theory of Special Relativity and wave mechanics. The result was stunning indeed; new quantum features were discovered called the 'spin', which arose naturally from the Dirac equation. Also, it was discovered that charged particles have their analogue with an opposite charge, known now as an anti-particle. This allowed an imagining of a whole universe consisting of anti-matter. When matter and anti-matter meet, a flash of light is created, causing both matter and the anti-matter to be annihilated. Moreover, a whole sea of negative energy states was discovered, which was thought to be the source of anti-particles. This was called the 'Dirac Sea.'

Again, physicists felt the need for a more general framework that could help describe the world in a more comprehensive way. Meanwhile, much water passed under the bridge. New problems arose because the new mechanics gave rise to different interpretations. Albert Einstein and Niels Bohr got into deep discussions during the 1930s, arguing about the meaning and implications of the new quantum mechanics. The question of indeterminism was at the heart of these discussions. Einstein held the view that the indeterminism shown by the quantum description of the microscopic world pointed to the incompleteness of the theory. He, along with several colleagues, devised a thought experiment demonstrating how quantum mechanics could lead to absurdities. These included having a pair of particles (a positron and an electron for example), which had been generated from the same source, becoming entangled, even if they were separated by light years of distance. In some respects, the discussion turned out to be philosophical, posing fundamental questions about the reality and objectivity of our knowledge. Indeed, some of these questions are still being discussed today. These questions need to be resolved if we are to obtain objective and satisfactory answers which will help us understand the meaning of our existence and destiny.

There were two options for the mathematical formulation of the physical theory: discrete structure and continuous differential fields. The second option won the most acceptance, since most physicists were familiar with continuous fields and differential calculus. The classical theory of fields achieved a great deal for physics. The sophisticated Hamiltonian and Lagrangian formulations of classical mechanics enabled precise solutions to be achieved for many problems, including planetary motions and describing, in detail, the effects of the planets and the Sun on each other. Such calculations enabled predicting the positions of celestial objects to a very high degree of accuracy and, therefore, predicting the time of lunar and solar eclipses up to fractions of a second.

In quantum electrodynamics, the notion of field and canonical formulation of the equation of motion, and the subsequent analysis of the physical problems in atomic physics, proved to be very successful indeed. This consolidated differential calculus with the continuum approach to formulating the structure of quantum field theory during the 1950s and 60s.

This success in the classical theory of fields pushed physicists to go for the option of continuity, describing quantum systems in terms of a continuous field and trying to obtain discreteness as a by-product of the position of these fields. This idea was very compelling indeed, but not free from problems. The new quantum fields approach achieved a great deal, starting with the quantisation of the Maxwell field. Julian Schwinger, Richard Feynman, Freeman Dyson, and others, were able to formalise a theory of quantum electrodynamics that achieved great theoretical success during the 1940s and 50s. This work became a firm paradigm for dealing with quantum fields and obtaining quantisation in particle physics theory. However, it was soon found that ugly infinities appeared when calculating some fundamental quantities, such as the charge and mass of the electron. These infinities, which stood as a stumbling block hindering the progress of particle physics theory, were identified as a "chronic disorder in the theory."

Once we adopt continuity as a basic structure, we ought to realise that we are embedding infinities into the foundation of our systems. Such infinities have to pop up in our calculations. Here, I don't mean to say that differential calculus is not suitable when dealing with quantum systems, surely not, but what I am saying is that the assumption of a field as a substratum for systems is, itself, a problem. Should we adopt some vision to take into consideration the discrete nature of systems, it would, perhaps, be of help in providing us with classical systems. Despite the great success achieved with Quantum Theory, there are still several outstanding problems in understanding its implications. These fundamental, conceptual problems are hindering the progress of contemporary theoretical physics.

NEW THEORIES

As unsolved problems became a burden on the shoulders of physicists, they were pushed into thinking further outside the box in order to find new approaches that could lead them to providing reasonable solutions. This resulted in considering new ideas, such as the motion of vibrating strings, to describe systems on minute scales. This idea is very beautiful, and one might expect to get much physics out of it, but unfortunately it was soon found that the theory went down several routes and lost its elegance and simplicity, as well as the power of predictability, which is an essential feature of modern scientific theory. For this reason, the proposal has been harshly criticised by some physicists.[115]

The question of quantum indeterminism seems to be an unreasonable result of quantum description. The probabilistic nature of events described by quantum mechanics can hardly be accepted in the framework of a rational philosophy that considers the universe to be a gigantic machine. It seems that

115 See for example Lie Smolin, *The Trouble with Physics.*

Western mentality, which is a continuation of Greek mentality, being accustomed to treating things in exact measures and according to fixed laws, finds it hard to accept the indeterministic character of events that might happen at any time, and with much uncertainty of their values. This is, perhaps, the reason why, soon after the problems of quantum field theory became evident, a deterministic approach was suggested by David Bohm in the early 1950s. Bohm tried to restore indeterminism by assuming some sort of hidden, and undetected, variables to be at play. The Bohm proposal is still not favoured amongst physicists, despite its recent promotion by some authors. The basic idea holds that quantum mechanics is an incomplete theory. This was conveyed by Einstein several times, when he expressed his rejection of the probabilistic nature of quantum systems, saying, "God does not play dice."

However, the fact that the local hidden-variable theories were eliminated by the verification of Bell's Theorem, with the remaining non-local hidden variable theories being eliminated by the verification of the Leggett's Theorem, and given that the Bohm Theory could not provide any experimental evidence for its superiority, makes one inclined to believe that the standard form of quantum mechanics is the best authorised presentation available on the scene now. This is despite the inherent problems in interpreting the outcome of the theory, including the problem of measurement.

With respect to the quantum measurement problem, we should be aware that, although, quantum indeterminism has now become an experimental fact, the interpretation of such indeterminism is still a theory, and that this is where the root of unhappiness lies. The available interpretation, which has its origins in the Copenhagen School, is not lucid enough to be accepted rationally. On the same footing, we should recognise that the Heisenberg Uncertainty Principle is a naturally inherent uncertainty when it comes to determining the values of some physical observables, which correspond to incompatible operators which work as generators of one another. So, I feel that there is no way we can get rid of this principle unless we exchange Quantum Theory and wave-mechanism with another, more fundamental, theory that replaces the very foundations of quantum mechanics and leads to a more general theory. This would be similar to the way that Newtonian mechanics and his law of gravity were replaced by the Theory of Relativity. For this purpose, we must change the whole structure of our physical theory and, perhaps, develop new physics.

ATOMISM AND CAUSALITY

The problem of causality and its position in *kalām* will be dealt with in Chapter Ten of this book. However, I find it necessary here to point out the relationship between the concept of causality and atomism according to *kalām*. Wolfson noticed that those who negated causality believed in atomism and those who negated atomism believed in natural causality. He says, "those who believed in atoms denied causality, whereas those who believed in infinite divisibility of matter affirmed causality."[116]

In fact, this position has a profound philosophical basis, for atomism, in describing things in discrete un-related parts (except when part of an aggregate), would surely require some agency to bind its parts and give it the status of being an interaction. This makes the events of the world, including all their properties, subject to the action of an agency other than their natural tendencies. This agency is assumed to be embodied in those parts. In fact, this view goes further towards being indeterministic, despite being ruled by certain laws which we recognise from the regularity of the events. This is the essence behind the denial of causal determinism. This topic will be further discussed in subsequent chapters.

116 Wolfson, 468

Chapter Four

Temporality

Temporality is a term which is directly related to time-dependence. A temporal object is that which is ephemeral. This implies that it is in a state of persistent change, thus implying mutability and, in some sense, it is the opposite of *eternity*.

According to the Mutakallimūn, the world was created, meaning that it had a start *along with* space and time.[117] The constituents of the world are said to be temporal (*muḥdath*), which means that they came into existence at a certain time. However, whether it was created out of something or out of nothing is an open question, but the important fact to realise is that it could have been created out of something belonging to the world itself, since such an action cannot be called creation but may well be called transformation.[118]

117 We do not say *starts in time* because, according to the Mutakallimūn, the creation of the world has been carried out not *in* time but *along with* time. Time before the creation of the world did not exist. After the first moment of creation, time existed and everything created had a start in time (see the chapter on space and time and the discussion of al-Ghazālī in Chapter Eleven.)

118 In *kalām* this question is discussed under the eternity and temporality of the world. Wolfson discusses whether the world has been created out of something or from nothing, taking into consideration some of the verses of the Qur'an. However, he could not reach a conclusive verdict (see: Wolfson, *kalām*, 356–376)

The world could have been created out of nothing (*ex nihilo*) or created out of some other stuff that was not part of the constituents of the material world itself. The only form that we know of out of which the world could have been created is the quantum vacuum. This is supposed to have existed forever; although it is a legitimate question for those who believe in Creation to claim that the quantum vacuum itself, as described by modern theories of physics, is in need of a Creator and Sustainer.

Al-Jurjānī defines the world as "everything of existent except God."[119] Al-Ghazālī has adopted the same definition in his books, notably, *al-Iqtiṣād fi al-I'tiqād* and *Tahāfut al-falāsifah*. In *Lisan al-Arab*, the famous Arabic dictionary, Ibn Manẓūr says, "the world is the whole of creation, and is said to be what the great heavenly orbit [The Atlas] contains."[120] This shows that the term world is a synonym for what we now call the universe. However, the Arabic expression for the universe is *kaūn*, and this word is a specific term in *kalām*, meaning "a name for what happens suddenly, such as the conversion of water suddenly became air in one shot. If [the change] happens gradually then it is the motion." Then he says, "and it was said that the *kaūn* is the occurrence of form in matter after non-existing."[121] However, I will be using the term *world* to express all the content we can see at all levels, that is, microscopic and macroscopic, including what is now called the whole universe.

Views About the Existence of the World

The current literature classifies views on the origin of the world as follows:[122]

1. Those who say that the world is old, that it never had a definite start with time and is a block closed in on itself. Those who hold this view negate any purpose for the existence of the world.
2. Those who believe that the world has been created by a spiritual agency. This is mostly the religious attitude which includes:

    a. Those who say that God has made the world in the same way as a carpenter would make a chair from wood. This means that there is a basic material for the creation. This attitude was adopted by Plato and his school of thought in Greece, and Abū Baker al-Rāzī in Islam.
    b. Those who believe that the world is eternal, but that God is eternal also, and that He acted as a final cause for the world. This attitude was held by Aristotle in Greece and Ibn Rushd in Islam.
    c. Those who believe that God created the world, not out of an eternal material, but as an emanation from himself. This emanation is thought

119 Al-Jūrjānī, al-Ta'rīfāt: al-'ālam

120 Ibn Manẓūr (1993). *Lisan al 'Arab*. Beirut: Dar Ṣadir, v.s: *'ālam*

121 Al-Jurjānī, al-Al-Ta'rīfat: kaūwn.

122 Al-Alusi, *Kalām*.

to be temporal in itself but eternal in existence. This attitude was held by Plotinus and the Muslim philosophers, al-Fārābī and Ibn Sīnā.

d. There are those who believe that God created the world out of nothing at a certain instance, which was determined by the will of God. This attitude of creation *ex nihilo* was held by most of the Mutakallimūn, Ibn Ḥazm and Al-Ghazālī.

The aim of this chapter is to present the views of the most prominent of the Mutakallimūn regarding the problem of creation, briefly explain their arguments in this respect, and then evaluate those views with reference to contemporary scientific knowledge about the existence of the world. I will concentrate on the main points of those arguments, irrespective of the historical context.

CREATION OUT OF NOTHING (*EX-NIHILO*)

The Mutakallimūn routinely begin with the Qur'an in their investigation of the world. This fact has been emphasised by prominent researchers, such as Richard Walzer and William Lane Craig, as cited in the Introduction of this book. The differences between the Mutakallimūn stem from their attitude towards interpreting the verses of the Qur'an. This is a vital point which one should take seriously when studying *Kalām*, i.e. that the different trends of thought originated from these different interpretations, of the verses of the Qur'ān. Creation has been mentioned in the Qur'ān many times. For example, we read in Surat al-An'ām:

> *And He it is Who created the heavens and the earth with truth. And when He says, Be, it is. His word is the truth and His is the kingdom on the day when the trumpet is blown. The Knower of the unseen and the seen; and He is the Wise, the Aware. (6:73)*

Also, in Surat al-A'rāf we read:

> *Surely your Lord is Allah, Who created the heavens and the earth in six periods, and He is established on the Throne of Power. He makes the night cover the day, which it pursues incessantly. And (He created) the sun and the moon, and the stars; made subservient by His command. Surely His is the creation and the command. Bless Allah, the Lord of the worlds! (7:54)*

And we read in Surat al-Anbiyya:

> *And He it is Who created the night and the day and the sun and the moon. All float in orbits. (21:33)*

> *Who made everything He has created good, and He began the creation of man from clay. (32:07)*

In Arabic lexicons, the meaning of the word *khalq* (creation) is producing something out of nothing. For example, al-Raghib Asfahānī says, "creation originally means to produce an object out of no origin and no previous example, and could be used to express producing something out of another, like the verse which says: You were created out of one soul."[123] In *Lisān al-Arab*, Ibn Manẓūr says, "creation in Arabic means producing something unprecedented, and everything created by God is what he has started unprecedented."[124]

This short survey, supported by a brief exposition from the Arabic, shows that the verses of the Qur'an do not offer much help in respect of telling us whether the creation of the world was out of something (*min shay*) or from nothing (*min la shay*). This is why interpreters of the Qur'an were confused when it came to understanding this verse from Surah Tur:

> *Were they created from nothing, or were they themselves the Creators? (52:35)*

They tried to interpret this verse to mean that the original creation was made out of an eternal matter.[125] However, as far as I can see, the above verse endorses the ability of the Creator through stressing His attributes, and indicates that creation cannot be either spontaneous or self-supporting.

This is what Zechariah (Prophet Zakaria) was reminded of by God telling him in Surah Maryam:

> *I have created you before and you were nothing. (19:09)*

It should be noted that this verse is not in contradiction with the verse from Surah at-Tariq (86:06) which says that man was created '*out of splashing water*', and the verse in Surah an-Nur (24:45) which says that God has '*created every animal out of water.*' These verses stress that those who are created were brought into being out of something and that they are temporal. Wolfson directs our attention to the verse in Surah Fussilat, which says:

> *Then He directed Himself to the Heaven and it was a smoke, so He said to it and to the Earth: Come both, willingly or unwillingly. They both said: We come willingly. (41:11)*

He takes this as evidence for saying that the creation started from smoke. This could be true if we consider the creation mentioned here to indicate a specific stage; that is creation of the solar system. However, understanding this verse depends on understanding what is meant by the term Heaven. In an extensive study, a colleague, al-Zu'bi, and I addressed this question. Our findings suggest

---

123 Al-Aṣfahānī, al-Rāghib *al-Mufradāt fi Gharīb al-Qur'ān*, Safwān Dawūdī (ed.) Damascus: Dar al-Qalam, (1412 A.H.), 296.

124 Ibn Manẓūr, Lisan: *khalq*

125 Wolfson, *Kalām*, 356.

that this term could have several meanings, one of which may be the seven celestial spheres or orbs, as suggested by Greek astronomers.[126]

There is no indication in the Qur'ān that the universe is eternal, but there are several verses which point to God's creation and re-creation being a continuous process. For example, in Surat Younis, we read:

> *Is there anyone among your associate-gods who produces the first creation, then reproduces it? Say: Allah produces the first creation, then He reproduces it. How are you then turned away! (10:34)*

And in Surat al-Naml we read:

> *Or, Who originates the creation, then reproduces it, and Who gives you sustenance from the heaven and the earth? Is there a god with Allah? Say: Bring your proof, if you are truthful. (27:64)*

In Surat al-Rūm we read:

> *Allah originates the creation, then reproduces it, then to Him you will be returned. (30:11)*

and

> *And He it is, Who originates the creation, then reproduces it, and it is very easy to Him. And His is the most exalted state in the heavens and the earth; and He is the Mighty, the Wise. (30:27)*

In these verses, the power of God is elaborated through the process of creation. However, we should take note that re-creation is taking place at different levels. In any case, what we understand from all this is that the process of creation is an open and complex question. The issues which arise here and in the theory of creation, according to Islamic *Kalām* is dealt with in Chapter Five and subsequent chapters of this book.

MOTIVATIONS FOR THE TEMPORALITY OF THE WORLD

The Mutakallimūn always maintain that eternity is the sole character of the Divinity. Therefore, they argued, it followed that any eternal object would share existence with the Divinity. This, of course, would be in sheer contradiction with the unity of God and the Islamic belief in *Tawḥīd*.

For the above reason, the Muʿtazilis refused to consider divine attributes as an addition to divine characters. Instead, they considered those attributes to be an essential part of the self-character of God (*Al-sifat 'ayn al-dhāt*). The Muʿtazilis thought that acknowledging divine attributes as being an addition

126 Altaie, M.B. and M. K. al- Zuʿbi, "The concept of Heaven and Heavens in the Qur'an and Modern Astronomy," *JJIS*, vol. 4, No.3, 223–249. 2008.

to the divine character would allow for characterising God as more than one and that this might lead to polytheism. In fact, this attitude of the Muʿtazilis has no explicit support in the Qur'ān, except in what can be taken as pointing to divine attributes as being a metaphor for expressing the power of God, His omniscience and omnipotence. However, it should be taken into consideration that whenever the Qur'an mentions divine attributes, a warning is given that these should not be linked to polytheism. Otherwise, no Muslim could confirm that God can hear with ears like ours or see with eyes like ours, or have a tongue to speak with, and so forth.

It is widely acknowledged in Islam that such corporeal attributes are not to be identified with the Divine, since the Qur'an clearly stipulates that, "*nothing resembles him.*" In any case, the Mutakallimūn found that the best argument for defending the unity, singularity and eternity of God is to believe that everything else is temporal, while Allah is the only eternal. This was the main motivation behind the insistence of the Mutakallimūn on the principle of temporality.

The Mutakallimūn consolidated their views through simple but profound reflections. They found that there were several, natural observations that indicated that the world and its content are temporal. For example, they noticed that motion occurs in the body when immobility disappears. This change in the state of a body is an indication of the temporality of motion. It is an emergent character that is added to the body which was originally at rest. We may thus consider that the main argument behind temporality is the *change*, since this is the direct evidence that things cannot be eternal.

This kind of argument has been used in the Qur'an in several verses. For example, Abraham, the father of Ishmael and Isaac, is said to have searched the sky looking for God. When he saw a star (or a planet) he said: this is my God, but when that object faded, he observed: this fading object cannot be my God; it fades. This we read in the following verses of Surah al-Anam:

> *So when the night overshadowed him, he saw a star. He said: Is this my Lord? So when it set, he said: I love not the setting ones. Then when he saw the moon rising, he said: Is this my Lord? So when it set, he said: If my Lord had not guided me, I should certainly be of the erring people. Then when he saw the sun rising, he said: Is this my Lord? Is this the greatest? So when it set, he said: O my people, I am clear of what you set up (with Allah). Surely I have turned myself, being upright, wholly to Him Who originated the heavens and the earth, and I am not of the polytheists. (6:76–79)*

Here we see that change in the state of celestial objects motivates Abraham to negate the divinity of such objects. This is an example of the simple arguments set by the Qur'an in order to direct the attention of mankind to the fact that, although inanimate objects like celestial bodies may look fascinating and beneficial, they cannot provide evidence to be God. The verses may also point to the historical development of man's religious faith.

Another example is from verses in Surah al-Baqara, which tell us a short story about the man who passed through a deserted village, asking the question of how God could re-create this village:

> *Or like that who passed by a town, and it had fallen in upon its roofs. He said: When will Allah give it life after its death? So Allah caused him to die for a hundred years, then raised him. He said: How long hast thou tarried? He said: I have tarried a day, or part of a day. He said: Nay, thou has tarried a hundred years; but look at thy food and drink—years have not passed over it! And look at thy ass! And that We may make thee a sign to people. And look at the bones, how We set them together then clothe them with flesh. So when it became clear to him, he said: I know that Allah is Possessor of power over all things. (2:259)*

These examples explain to Muslims that change indicates management by some agency, which is effecting those changes. In the same context comes the argument that continued re-creation allows for a more generic argument of change, which makes the need for God something indispensable.

Philosopher's Motivation for the Perpetuity of the World

Why should philosophers believe in the eternity of the world? Why should they deny that the world could have a point in time when it came into existence?

First of all, philosophers understand that time has always existed as there is no reason why it should not exist. Time in the philosophy of Aristotle, for example, is an absolute flow that has no beginning, so it will be hard to imagine that the world had a start in time. Also, philosophers ask why the world would have been created at a specific moment of time? No sensible answer can be found for such a question. However, once we realise that, originally, the world was not created *in* time but *with* time, then we understand that it could have a beginning, since time, itself, would then have a beginning, which was the commencement of change. Before creation there was no change; time did not yet exist. But once time was created, everything began to exist and had to obey the law of change. Here, the most fundamental philosophical argument about the eternity of the world breaks down.

But could it be that change itself is an eternal law of existence? By such an argument we would have an eternal world of continuous changes. This might be theoretically possible, of course, but it should not break with the argument of the temporality of the constituents of the world, as every object in the world is subject to change and is, therefore, temporal. However, since the world is a collection of these temporal objects, it follows that the whole world may be considered to be under change and, therefore, temporal despite the rationale that change itself is eternal. In any case, it remains a matter of scientific investigation to search for the beginning of the whole universe. Although, as we will see shortly in this chapter, such a question might be treated philosophically with different interpretations.

There will always be the question of the eternity of the world as long as change can be seen to apply to everything. However, as demonstrated above, temporality means change and the eternity of change does not imply the eternity of the world necessarily, as this would mean that the world renews itself every time a change of its constituents happens. This occurs, for example, in the case of a cyclic universe, which is re-created after every cycle.

ARGUMENTS FOR TEMPORALITY

The Mutakallimūn follow two approaches for arguing for the temporality of the world: the first is to negate the eternity of the world, and the second is to prove the creation of atoms (the indivisible part) and all the constituents of the world.

There were a number of Muslim theologians who considered the question of the temporality of the world seriously. They presented their arguments in the style of comparative studies, bringing in counter arguments or the views of other philosophers on the same topic. Two such theologians were Ibn Ḥazm and al-Ghazālī. It is true that neither was a formal Mutakallim but in their argument they followed the approach of *kalām*.

The Arguments of Ibn Ḥazm

Ibn Ḥazm devotes a large section of his famous book, *al-fiṣal fi al-milal wa alhwā' wa al-niḥa*, to discussing the views of Dahrīya (the eternalists) who said that the world is eternally as it is and was never created. Ibn Ḥazm has two main objections to this. These are:

1. that the creation of the world out of nothing is not self-evident.
2. that the action of creation itself is either:

   a. associated with the Creator naturally, whereby he has no actual intervention of will, or
   b. that he has created the world wilfully in order to, either achieve good, or to eliminate bad. In such a case, the Creator would be bound to do good or eliminate bad in order to achieve his goal, and as this applies to temporal beings then, consequently, the Creator needs a Creator himself.

After reviewing these basic arguments of the eternalists (Dahrīya), Ibn Ḥazm goes on to present his counter arguments as follows:

(1) **the argument of finiteness:** this argument, in short, says that the world, as we see it and comprehend through our senses, is finite in size (in its spatial extension), as well as in time (age), as it can be demonstrated to have a beginning. Ibn Ḥazm says, "the world is finite and has a beginning inevitably."[127] Furthermore, he states that, "the world actually exists and is bound by numbers and countable by nature, so it is finite, all the world is finite." Here Ibn Ḥazm points to the Qur'an and quotes the verse, which reads *'and everything has a magnitude.'* (13:08)

In order to demonstrate the finiteness of the world, Ibn Ḥazm says whatever has an end (has ended) cannot be increased since "the meaning to increase is to add to the finite something of its kind increasing its content or area."[128] Here, Ibn Ḥazm discusses time in terms of something that has lapsed since the first creation took place, using astronomical time as a unit. He takes the time that Saturn takes to orbit the sun as an example and says that this planet is known to cover the Zodiac in 30 years approximately, so when Saturn makes one trip through the Zodiac, i.e. the great orbit it has made eleven thousand, less fifty, periods, and, undoubtedly, the eleven thousand, less fifty, is more than one period. Therefore, he concludes, "that if the world is eternal then one infinite (counted in terms of Zodiac period) is more than the other infinite (counted in terms of Saturn's period) by eleven thousand times or so, and this is impossible."[129]

In fact, Ibn Ḥazm obtained the figure of eleven thousand, less fifty, by multiplying 30 by 365; this is the number of days in 30 years, during which time the Zodiac makes 10,950 revolutions. So here he argues "that since the time units by which we can measure the age of the world is finite, therefore, on using different units for the measure, we will get different numbers for infinity, accordingly; this shows an inconsistency in the measurements which cannot be resolved." With this argument, Ibn Ḥazm wishes to refute the claim that the universe is of an infinite age. Obviously, this argument assumes that the infinite is a countable figure, which itself is a self-inconsistency. In any case, it remains a problem that needs to be resolved.

The point to make here is that, in spite of the fact that Ibn Ḥazm has negated, elsewhere, *kalām* atomism, we find him here talking in terms of discrete parts and considering the whole as being the sum of these parts. This is quite different from Aristotle and Ibn Rushd's understanding of the whole, where they consider it as one unit (see the arguments of Ibn Rushd against

127 Ibn Ḥazm, Fisal, 1/20
128 Ibid.
129 Ibid.

the indivisible part of *kalām* which are presented in a foregoing section). Ibn Ḥazm says, "and the meaning of a part is to be part of a whole and the whole is the sum of all parts. As such [the term] whole and [the term] part comes in describing every divisible whole."[130]

(2) **The argument for the integrity of space and time:** Ibn Ḥazm analyses time to assert the connection between space, time, and the body. This view will be discussed in more detail when I consider the concepts of space and time in *kalām*. Here, I can say that the argument for the integrity of space and time establishes the understanding that space and time are not absolute but are relative, measure-dependent, and that space, time, and the body form one combination. Consequently, Ibn Ḥazm negates the existence of infinite space as well as infinite time. Additionally, Ibn Ḥazm explains that space and time were born along with the world, and that the Creator does not exist in time or space, and that He is neither at rest nor in motion. Also, Ibn Ḥazm asserts that the world cannot be infinite since the Qur'an has affirmed the measurability of the constituents of the world.

It is notable that, in the above two arguments, Ibn Ḥazm is using two principles of *Daqīq* al-*kalām*: the principle of discreteness (but not necessarily atomism), and the principle of the integrity of space and time. Incidentally, his refusal to acknowledge the indivisible part of *kalām* was based on his argument that such a part does not have ontological status, as shown before. This fits well with the methodology of Ibn Ḥazm in defining what is acknowledged to exist. The future is not a tangible thing for him, as long as it has not come yet. He says, "and whatever has not yet been in time or a realized item, or an accident then it is not to be considered a thing and is uncountable or stands as an end, and should not be described as a thing because it does not exist."[131]

It might be interesting also to learn that Ibn Ḥazm negated the idea that the world could have created itself. However, the condensed argument he provides for this requires much elaboration which, I regret, is beyond the scope of the present chapter. Also, of interest is that Ibn Ḥazm used the argument of design to support the need for a Creator and, subsequently, to assert that the world is temporal.

Al-Bāqillānī on Creation

In his book, *Tamhīd al-awail wa talkhīs al dala'il,* al-Bāqillānī used *kalām* atomism to demonstrate that the world is temporal, saying, "all the world, the lower and the upper, is composed of two entities, I mean the *jawhar* and the accidents, and it is all temporal. The evidence of its temporality is what we

130 Ibn Ḥazm, Fisal, 1/22.

131 Fisal, 1/24.

have shown previously of the demonstration about accidents."[132] He then goes on to state that, "the accidents are temporal and the evidence for their temporality is the vanishing of motion upon rest, because if it did not vanish when rest occurs then both would be existing in the body together and this means that the body will be moving and at rest altogether which is necessarily null."[133]

Al-Bāqillānī then proceeds to prove the existence of God by using direct rhetorical arguments in saying that the Creator, being alive and omnipotent, negates the notion that the created is able to create itself. This indicates that al-Bāqillānī does not accept the spontaneous self-creation of the world out of nothing, because the world is neither alive, nor is it omnipotent. Then he says that the Creator of the objects cannot be similar to it because should this be so, then they would either share eternity with Him or He would share their temporality. Then al-Bāqillānī asserts that the Creator should be one and not many, since if there were many, the world would be corrupted. For this he quotes the Qu'ranic verse in Surah al-Anbiyya (21:22) which says, "*say if there were gods except Allah it would have corrupted.*" He also asserts that the Creator cannot be created using arguments similar to those used by Ibn Ḥazm. Then, al-Bāqillānī moves on to discuss the purpose of creating the world, a matter I will discuss later in this chapter.

Al-Ghazālī on Creation

Al-Ghazālī extensively discussed the arguments of the philosophers with regards to the question of creation and eternity of the world. In his book, *Tahāfut al-Falāsifah*[134] (The Incoherence of the Philosophers), Abū Hamid al-Ghazālī uses some of the arguments that were originally used by Ibn Ḥazm; for example, the argument in which the periods of celestial bodies are given. However, al-Ghazālī expanded these arguments by saying that the eternity of the world requires that the planets make an infinite number of revolutions through the Zodiac, but such a state is by no means self-evident as it contradicts the fact that such periods have a half and a third and a quarter. He is trying to use the fact that periods can be differentiated, or partitioned, in order to negate their infinite extension.

Also, al-Ghazālī uses the argument of the possibility that the total number of periods, made by any celestial object, can be even or odd, and that it cannot be infinite since infinity cannot be even or odd. Certainly, the question of infinity is something rather subtle since such a concept was, and still is, one of the most confusing concepts in science. Therefore, one should admit that

132 Al-Bāqillānī, *Tamhīd*, 41.

133 Ibid.

134 Abū Hamid al-Ghazālī, *The Incoherence of the Philosophers*, Edited by Michael Marmura, Utah: Brigham Young University Press, 2000.

such rhetorical arguments have a shallow intellectual content and are of low scientific value.

About the Particularizer and the Particularization

Al-Ghazālī confronted the argument of the philosopher which stated that there was no reason to create the universe at a specifically chosen time, with a specific shape and designated properties, and, which stipulated that the universe must be eternal. Al-Ghazālī responded to this by saying that a "particularizer" (*murajeh*) would be required for such a change to happen. This particularizer is the will of God. Then he goes on to discuss the question of "before" and "after" in the context of discussing the chosen moment for the creation of the world, where he brilliantly shows that these two terms have no absolute meaning, similar to "above" and "below", which are designated with respect to a certain place.

Al-Ghazālī defines the will as "an attribute which is needed to distinguish something from another thing that is identical to it." This means that the particularizer distinguishes between two things that have the same properties. For example, a person finds two dates in front of him, and he picks up one of them but not the other; he does so according to his own will, thereby indicating that he has played the role of the particularizer. The concept of the particularizer is very important as well as subtle, because it is closely related to the will and is expressive of an implicit property which enables one of two equally probable events to occur. Al-Ghazālī understands that something should favour the occurrence of one particular event rather than the other, and this is the will. This concept of the particularizer, in the context of the creation of the world, provides an answer to the question: why should there be something rather than nothing?

On the question of the timing of creation, al-Ghazālī exposes the philosopher's question which asks, "If this were allowed, then it would be permissible for the world, whose existence and nonexistence are equally possible, to originate in time, and the side of existence, similar in terms of possibility to the side of nonexistence, would thus be specified with existence without there being that which would specify [it]."[135] In addressing the philosopher's argument, al-Ghazālī says, "For if the eternal's specific relation to one of the two [equally] possible [existents] through [sheer] coincidence is allowed, then the uttermost unlikely thing would be to say that the world is specified with specific shapes when it is [equally] possible for it to have other shapes instead; whereupon one would then [also] say that this occurred in a manner coincidentally, in the same way that you [theologians] have said that the will has specifically related

135 Al-Ghazālī, *Incoherence*, 21.

to one time rather than another and one shape rather than another by coincidence. If you then say that this question is superfluous because it can refer to anything the Creator wills and reverts to anything He decrees, we say, "No! On the contrary, this question is necessary because it recurs at all times and attaches to those who oppose us with every supposition [they make]."[136]

In further response to these arguments of the philosophers al-Ghazālī says:

> [To this] we say: The world came to existence whence it did, having the description with which it came to exist, and in the place in which it came to exist, through will. Will being an attribute, whose function is to differentiate a thing from its similar. If this were not its function, then power would be sufficient. But since the relation of power to two contraries is the same and there was an inescapable need for a specifying [agent] that would specify one thing from its similar, it was said: the Eternal has, beyond power, an attribute that has as its function the specifying of one thing from its similar. Hence someone's statement, 'Why did the will specifically relate to one of the two similars?' is akin to the statement, 'Why does knowledge entail, as a requirement, the encompassing of the object of knowledge as it is?' For [to the latter] one would reply, 'This is because knowledge stands as an expression for an attribute that has this as a function.' Similarly, 'Will' stands as an expression for an attribute whose function—nay, its essence—is to differentiate a thing from its similar.[137]

Then al-Ghazālī argues on behalf of the philosophers saying:

> [To this the philosophers might then] say: Affirming an attribute whose function is to differentiate a thing from its similar is incomprehensible - indeed, contradictory. For to be similar means to be indiscernible and to be discernible means that it is dissimilar. One should not think that two [instances of] blackness in two places are similar in every respect. For one is in one place, the other in another. And this necessitates a differentiation. Nor are two [instances] of blackness in the same place at two different times absolutely similar. For one differed from the other in terms of time, how could it be similar to it in every respect? If we say that the two [instances] of blackness are [two] things similar to each other, we mean by it [similar] in blackness related [to the two instances] in a special, not in an unrestricted, sense. Otherwise, if place and time are unified and no otherness remains, then neither the two [instances] of blackness, nor basically duality, are conceivable. This is shown to be true by [the fact] that the expression 'will' [as applied to God] is a borrowing from our 'will.' It is inconceivable of us that we would differentiate through will one thing from its similar. Indeed, if in front of a thirsty person there are two glasses

136 Ibid.
137 Ibid., 22.

> of water that are similar in every respect in relation to his purpose [of wanting to drink], it would be impossible for him to take either. Rather, he would take that which he would deem better, lighter, closer to his right side, if his habit was to move the right hand, or some such cause, whether hidden or manifest. Otherwise, differentiating something from its like is in no circumstance conceivable.[138]

Here, in response to this argument, al-Ghazālī says:

> The objection [to this argument of the philosophers] is twofold: the first is regarding your statement that this is inconceivable: do you know this through [rational] necessity or through theoretical reflection? It is impossible [for you] to appeal to either of these. Moreover, your using our will as an example constitutes a false analogy that parallels the analogy [between human and divine] knowledge. God's knowledge differs from human knowledge in matters we have [already] established. Why, then, should the difference between [the divine and the human] in the case of the will be unlikely? Rather, this is akin to someone's statement: 'An essence existing neither outside nor inside the world, being neither connected nor disconnected with it, is inconceivable because we cannot conceive it on our own terms,' to which it would then be said, 'This is the work of your estimative faculty -rational proof has led rational people to believe this.' With what [argument] would you then deny one who says that rational proof has led to the establishing of an attribute belonging to God, exalted be He, whose function is to differentiate a thing from its similar? If the term "will" does not correspond [to this attribute], then let it be given another name; for there need be no dispute about names.

As for the argument of the philosophers where they say in the previous paragraph that, "otherwise, differentiating something from its like is in no circumstances conceivable", al-Ghazālī responds by saying:

> Even so, in our [own human] case, we do not concede that [the choice between similar things] is inconceivable. For we will suppose that there are two equal dates in front of someone gazing longingly at them, unable, however, to take both together. He will, inevitably, take one of them through an attribute whose function is to render a thing specific, [differentiating it] from its like. All the specifying things you have mentioned by way of goodness, proximity and ease of taking, we can suppose to be absent, the possibility of taking [one of the two] yet remaining. You are hence left between two alternatives. You could either say that equality, in relation to the individual's purpose, is utterly inconceivable, which is sheer foolishness,

138 Ibid., 22–23.

> the supposition [of this equality] being possible; or else, that if the equality is supposed, the man yearning [for the dates] would ever remain undecided, looking at them but taking neither through pure will and choice that [according to you] are dissociated from the objective [of taking a specific one]. This also is impossible, its falsity known by [rational] necessity. It is, hence, inescapable for anyone engaged in theoretical reflection on the true nature of the voluntary act, whether in the realm of the observable or the unseen, but to affirm the existence of an attribute whose function is to render one thing specifically distinct from its similar.[139]

In refuting the arguments of the philosophers, and in continuing his own argument, al-Ghazālī, says:

> The second way of objecting is for us to say: You, in your own doctrine, have not been able to dispense with the rendering one of two similar specifically [distinct], for [you hold] the world to have come into being through its necessitating cause, having specific configurations similar to their opposites. Why has it been specified with [certain] aspects [and not others], when the impossibility of differentiating one thing from its similar [as you uphold] does not differ, [whether] in the [voluntary] act or in that which follows by nature or by necessity?[140]

Following this discussion, al-Ghazālī goes into a greater discussion on the creation of the world by giving examples of the geometrical symmetry of the celestial sphere and the definition of the celestial poles. It is a very rich and wealthy intellectual exercise which reflects al-Ghazālī's deep understanding of geometry and astronomy, whereby he shows that the globe is symmetrical and the definition of position is only relative to a certain reference point, and is not an absolute.

In this part of the discussion, al-Ghazālī implicitly proposes that creation is a sort of broken symmetry that is done through the particularizer, which is the divine will. This part of the *Tahāfut* is a magnificent presentation of a highly intellectual argument and can be compared to contemporary discussions about symmetry breaking and spontaneous symmetry breaking, which accompany the process of creation from a out of a vacuum. This aspect of the argument for creation, as presented in the *Tahāfut*, is continued by other arguments from the relative designation of space and time where it is shown that al-Ghazālī treated time on an equal footing with space. This, I will present in the forthcoming chapters of Part Two, since it is part of a whole system of thought and opinion that was being discussed among the Mutakallimūn.

---

139 Ibid., 23–24

140 Ibid.

AL-GHAZĀLĪ ON THE SYMMETRY OF SPACE AND TIME

In addressing the problem of time, and the claim that God preceded the world not with time but with essence, al-Ghazālī says:

> Time is originated and created, and before it there was no time at all. We mean by our statement that God is prior to the world and time, that He was and there was no world, and that then He was and with Him was the world. The meaning of our statement, 'He was and there was no world,' is only [the affirmation of] the existence of the Creator's essence and nonexistence of the world's essence. And the meaning of our statement, 'He was and with Him was the world,' is only [the affirmation of] the existence of two essences. Thus, by priority, we only mean the appropriation of existence to Himself alone, the world being like an individual. If, for example, we said, 'God was and there was no Jesus, and then He was and Jesus with Him,' the utterance would not entail anything other than the existence of an essence and the nonexistence of an essence, then the existence of two essences. From this, the supposition of a third thing is not necessary, even though the estimative faculty does not refrain from supposing a third thing. But one must not heed the errors of estimative thoughts.[141]

Al-Ghazālī then goes on to discuss the possibility of God preceding the world by essence but not time, as proposed by the philosophers, where he presents another novel idea by pointing to the relative status of time and dealing with it on an equal footing with space. Here, we see that al-Ghazālī has comprehended the similarity and relativity of space and time. He discusses the designation of the before and after, saying:

> the basic thing understood by the two expressions is the existence of an essence and the nonexistence of an essence. The third thing, by virtue of which there is a difference between the two expressions, is a relation necessary with respect to us [only]. The proof of this is that, if we suppose the nonexistence of the world in the future and suppose for it another existence, then we would say, "God was and the world was not," this statement being true regardless of whether we intend by it the first nonexistence or the second nonexistence which is after existence. The sign that this is relative is that the future itself can become a past and is expressed in the past tense. All this is due to the inability of the estimative [faculty] to comprehend an existence that has a beginning, except by supposing a 'before' for it. This 'before,' from which the estimation does not detach itself, is believed to be a thing realised, existing—namely, time. This is similar to the inability of the estimation to suppose the finitude of body overhead, for example, except in terms of a surface that has an above, thereby imagining that beyond the world there is no place; either

141 Ibid. 31.

> filled or void. Thus, if it is said that there is no 'above' above the surface of the world and no distance more distant than it, the estimation holds back from acquiescing to it, just as if it is said that before the world's existence there is no 'before' which is realised in existence, [and the estimation] shies away from accepting it.[142]

Then he says:

> [Now,] one may hold estimation to be false in its supposition that above the world there is a void, namely, an infinite extension-by saying of it, "The void is in itself incomprehensible." As regards extension, it is a concomitant of a body whose dimensions are extended. If the body is finite, the extension which is its concomitant is finite and the filled space terminates [with the surface of the world], whereas the void is incomprehensible. It is thus established that beyond the world there is neither a void nor a filled space, even though the estimation does not acquiesce to accepting [this]. Similarly, it will be said that just as spatial extension is a concomitant of the body, temporal extension is a concomitant of motion. And just as the proof for the finitude of the dimensions of the body prohibits affirming a spatial dimension beyond it, the proof for the finitude of motion at both ends prohibits affirming a temporal extension before it, even though the estimation clings to imagining it and its supposing it, and not desisting from [this]. There is no difference between temporal extension that, in relation [to us], divides verbally into 'before' and 'after' and spatial extension that, in relation [to us], divides into 'above' and 'below.' If, then, it is legitimate to affirm an 'above' that has no above, it is legitimate to affirm a 'before' that has no real before, except an estimative imaginary [one] as with the 'above.' This is a necessary consequence. Let it then be contemplated. For they agreed that beyond the world there is neither void nor filled space.[143]

In this paragraph we see that al-Ghazālī has named time to be an extension in exactly the same way as calling space an extension. This is a huge step in understanding the similarity between space and time, as well as the relative measure of each of them, as seen by different observers from different positions. However, this designation should not be confused with the relativity of time as presented by Einstein's Theory of Special Relativity. The two things are different, though not unrelated. What makes this of real intellectual value is the proposition put forward by al-Ghazālī that beyond the world there would be neither a filling nor a void. But this is another matter that I will present in Part Three of this book.

---

142 Ibid. 32

143 Ibid. 33.

THE QUESTION OF PURPOSE

We should be careful how we approach the question of purpose, which is different from asking: does the world have a purpose? The first question is meant to ask whether God created the world for a need related to him personally, or whether it is a desire that was pressing on him to do so, while the second question asks whether the world has a purpose of its own, which is related to its content and the intelligence that it contains.

The above two questions are fundamentally different. In respect of the first question, the Mutakallimūn answered: No, God did not create the world for the purpose of personal need or desire, or benefit. This is indicated in Surah Al-Imran, "*And whoever disbelieves, surely Allah is above need of the worlds.*" (3:97).

Views of the Muʿtazilis

The Mutakallimūn dealt with the question of purpose early on in their formation. Al-Ashʿari presented it in his *Maqālat* briefly, pointing to what Abū al-Hudhayl had said that "God created the world for benefit of the people."[144] He also presented the view of al-Naẓẓām that the creation of the world was not for a reason but that the reason (*ʿilla*) is the purpose. Al-Ashʿari also notes that ʿAbad Ibn Sulaymān said that God has created His creation for a reason (*ʿilla*). However, other sources claim that the Muʿtazilis had much more to say about this question.

The View of Ibn Ḥazm

Ibn Ḥazm comments on the views of some of the Muʿtazilis without naming them. As for his own attitude, we find him saying, "we say that there is no reason for construction of God."[145] In another place he says, "our view which has been mentioned in other places, is that God does not act for a cause, but He does what he wills and whatever He does is just and wise."[146]

Views of Al-Bāqillānī

Under the title, "Allah made the world not for a purpose," al-Bāqillānī presented his opinions on this question and denied that "God has made the world for a reason which caused him to do it or a motive that caused his motion, or an emanation of some sort, or a being distracted by a cause, etc."[147] He sees these terms as unsuitable for describing Almighty God. However, I may say that the denial of al-Bāqillānī is not adequate. One could agree with all the reasons put forward by al-Bāqillānī for rejecting the need of a purpose and still ask the question: did God create the world for a purpose not known to us?

---

144 Al-Ashʿari, *Maqālāt al-Islāmiyyīn*, 107.

145 Ibn Ḥazm, *Fisal*, 18.

146 Ibid. 27

147 Al-Bāqillānī, *Tamhīd*, 50.

View of al-Shahrustānī

Under the title, "Refuting the reason and purpose in acts of God",[148] al-Shahrustānī wrote a long essay that does not seem to have a focal point except to negate the existence of a purpose in the acts of God. He contradicted the claim of some Muʿtazilis who said that the creation of the world was done as a favour to the world. Al-Shahrustānī wrote, "What does it mean to have God's purpose, and what does the good mean? You seem to understand that the purpose is to bring some benefit or prevent some bad events. The omnipotent is able to do everything and He has no need for anything."[149]

Unfortunately, I cannot see a solid argument with al-Shahrustānī that convinces me of his attempts to negate purpose and reason, when he could have considered taking an attitude similar to that of al-Bāqillānī instead. The Qur'ān tells us that Allah created death and life in order to test us, by doing good or bad deeds, and to assess our acts (Surah al-Mulk, 57:2). Those who do well are going to be rewarded and those who do badly are going to be punished. This is a good enough reason, I feel, for creation.

View of Al-Ghazālī

The question of purpose was not considered by al-Ghazālī, as far as I know, except through his discussion of the particularizer on the question of why the world was created, where he shows that it is the will of God which produced the world. We can say no more about this except that al-Ghazālī seems to have acknowledged the Ashʿaris' views, which stipulated that there is no purpose for the world other than God's own will.

A View About the Purpose

Primarily, we should differentiate between purpose and reason. The purpose is what something is aimed at, i.e. it is the target. The reason is the motivation behind taking a decision or doing an action. We might say: *what is the reason for going to the pharmacy*, and the answer is *to buy medicine*. We do not say: *what is the cause for going to the pharmacy*, but we may ask: *what is the cause of illness*, and the answer could be: *a viral infection*. However, the reason (*ʿilla*) includes some meaning of the conditions involved in the event. Al-Ashʿari presented different views about the *ʿilla* in his *Maqālāt*. In this presentation we know that the *ʿilla* should accompany the effect, otherwise it would become the purpose.

Perhaps we cannot see the actual reason for creating the world as it is, with all the good and the bad it contains, because our comprehension is still limited

148 Al-Sharustānī, *Iqdām*, 225–236.
149 Ibid. 227–228.

and not mature enough to understand the purpose for our existence. Nevertheless, reflecting on our world with our complex and advanced comprehension and consciousness, indicates that there must be a purpose for our creation.

Now, if we are following the original approach of the Mutakallimūn concerning the deduction of views and arguments, we should look at what the Qur'an says about the purpose of creation. First, we note that the world is said to have been created in order to make possible the existence of mankind on a planet such as Earth, equipped with all things necessary to make life possible. See, for example, the Qur'anic Surahs and verses: 13:03, 16:15, 20:53, 43:10. Secondly, the Qur'an says that everything in the Heavens and on Earth is made for the benefit and service of mankind. See, for example, the Qur'anic Surahs and verses: 14:32–33, 16:12–14, 39:05, 45:12–13.

Our comprehension is advancing, and once man realises that this universe is not purposeless, he will be able to recognise the true signs which will, in turn, lead him towards the ultimate reason for his creation. So, what can we expect for the future of our nature and our existence? The answer could be to reach an ultimatum of sharing life in this universe. Unfortunately, though, it seems we cannot physically cross the barrier of the spacetime we are living in.

We seem to be living in an unlimited extension of worlds, which we are glimpsing at thanks to our advances in physics. The question remaining is whether our consciousness could evolve to produce some kind of probe into the other worlds? This is uncertain. Nevertheless, we can surmise that such a development could be possible, and might be attained through a proper spiritual experience. We are not a sort of compact disk that will become worthless once it is chipped into pieces, since there is the possibility that we might be an advanced holographic plate which allows reconstruction of the content, even if it is chipped. The hologram theory of Leonardo Susskind allows for such reflections, so why is it that we don't have good faith, and believe?

CONTEMPORARY VIEWS AND OPINIONS

In this section I will present some of the prominent suggestions of contemporary scholars who have dealt with the question of creation both on the scientific and philosophical level. First, I will present the *Kalām* Cosmological Argument of William Lane Craig and then will discuss the objections to it.

The Kalām Cosmological Argument

William Lane Craig re-devised the *kalām* argument about temporality and creation, which was put forward by medieval Islamic scholars as proof of the existence of the Creator. Lane re-named it, "The *Kalām* Cosmological Argument" (KCA), a term he basically took from a statement made by al-Ghazālī wherein he says, "every being which begins has a cause for its beginning, and the world is a being which begins, therefore, it processes a cause for its beginning."[150]

Craig articulated this argument using two premises and one conclusion. These are:

1. Everything that begins to exist has a cause of it existence.
2. The universe began to exist.
3. Therefore, the universe has a cause for its existence.

Craig took benefit of the scientific discovery that the universe came into existence 13.7 billion year ago as proposed by the Big Bang Theory and backed up by astronomical measurements. Accordingly, his second premise found strong practical support from astronomical evidence. Consequently, Craig defended this *kalām* cosmological argument enthusiastically. However, there are objections to the KCA by those philosophers and scientists who do not find this argument to be fully supported by practical demonstration as well as lack of theoretical construct of the premises. Also, there were some scientific objections as to whether the universe really began to exist in the first place. Here I will give a brief overview of these objections and questions.

Objections of Quentin Smith

Quentin Smith is an American philosopher who worked on philosophical problems in physics and religion. Smith uses the argument, held by some theoretical physicists, to claim that the universe could have been created out of a quantum vacuum uncaused by converting the random vacuum fluctuation, which is thought to happen spontaneously without the need for a Creator.[151] This view has been championed in public debates by physicists such as Stephen Hawking and Lawrence Krauss.

However, the claim that quantum fluctuation (virtual state) can be converted into real particles is possible only when a gravitational field (in the form of curved space time) exists. Such a field (or curved background) is necessary to prolong the duration of the virtual particles and provide the necessary positive energy required by the real particle-antiparticle pairs to exist. This requirement was recently recognized by Stephen Hawking who declared that the universe can create itself with the aid of gravity. Hawking says, "because there is a law such as gravity, the universe can and will create itself from nothing."[152] The question is: where does gravity come from? Unless we know the answer to this question the issue will remain unfinished.

150 Al-Ghazālī, *al-Iqtiṣād*, 10.

151 Smith, Quentin. "The uncaused beginning of the Universe", *Philosophy of science* 55: 39–57, 1988; "Can everything come to be without cause?" *Dialogue* 33: 313–323, 1994.

152 Hawking, S. and Mlodinow, *The Grand Design* (2010). New York: Bantam Books, 10.

Grünbaum's Objections

Adolf Grünbaum is a prominent American philosopher of science at the University of Pittsburg in the USA. His argument against the KCA perhaps is strongest among the philosophical ones. Grünbaum's basic objection is that the first moment of the Big Bang when the universe was created does not qualify as a physical *event*. A physical event, according to Grünbaum, actualizes in time, that is to say it happens in time and marks a moment of beginning for that event. But since time started with the Big Bang itself, we cannot identify a moment that preceded it which would qualify as an event by itself. Consequently, Grünbaum denies that the universe beginning to exist as an *event* took place at the first moment of the Big Bang.

Obviously, we can acknowledge that the world has existed for some time, but from the philosophical point of view such an acknowledgment may lack recognition of the first moment of the existence as an *event*. However, by all means the universe began to exist, not necessarily at a well-defined moment, but certainly such a beginning is marked on the scale of time by t=0. This means that we cannot claim that there is no start for the universe, but we can only say that such a start is not well-understood now.

Besides this, I should mention that, physically, the initial conditions for the existence of the universe is part of the big question in physics; it deals with the initial conditions which are unknown, and since the laws of physics cannot probe the first moment, this problem will remain a challenge for physicists as long as gravity and quantum mechanics are not brought to terms. Craig responded to Grünbaum's objections with an article published in the British journal for the philosophy of science.[153]

Hawking and Hartle's No Start Argument

The famous theoretical physicist, Stephen Hawking, in collaboration with William Hartle, put forward a theorem that shows that the universe could have existed for an endless imaginary time before the Big Bang.[154] Some philosophers who were opponents of the KCA welcomed this proposition, considering it as disproof of the Big Bang and the invalidation of the beginning of the universe.

The Hartle-Hawking proposition was well-received by some people who found it supported their views of eternal universe, while others criticized the approach. In response to this, the standard model of cosmology (the Big Bang

153 Craig, William Lane. "The origin and creation of the universe: A reply to Adolf Grünbaum." *British Journal for the Philosophy of Science* (1992): 233–240.

154 Hartle, James B., and Stephen W. Hawking. "Wave function of the universe." *Physical Review* D 28.12 (1983): 2960.

Theory) was amended to include a phase of inflation at the very early stages of the development of the universe. The work was conducted by Alan Guth and his co-workers, who also coined the term 'inflation theory.' Despite the fact that this theory solved some fundamental problems with the standard Big Bang, it is still uncertain whether the scenario proposed is true. No solid evidence has been found to support a firm conclusion that inflation did take place.

Some alternative models, which are different from the standard Big Bang model, are also available. One of these is free from the problems inherent with the Big Bang Theory and provides a model for the universe without inflation. It suggests that the universe started from a Planck-size bubble inflating naturally by continuous creation of matter and energy to the present state.[155] This model also proposes that the universe was spatially flat naturally from the very beginning as its average density is always fixed to be equal to critical density.

The Hawking-Hartle theorem does not refute the real beginning of the universe since the claim that the universe could have existed in an infinite *imaginary* time is realistically meaningless as the imaginary time is not a *physical* time, it cannot be measured. Therefore, the Big Bang remains a real concept.

Weinberg's Purposeless Universe

Steven Weinberg is an eminent theoretical physicist, awarded the Nobel Prize in Physics for his contribution to unification of weak nuclear force and electromagnetic force. He is known for his eloquent style of scientific writing and has several specialized, as well as popular, scientific books.

Weinberg believed that there is no purpose for the universe and is unwilling to acknowledge any truth beyond what science can demonstrate. Weinberg says "the more the universe seems comprehensible, the more it is also seems pointless."[156] Weinberg does not deny that God may exist, but he cannot see that such existence could solve any problem. What Weinberg says might be true, if we limit the whole existence to the materialistic and physical world that we are dealing with. Then, it could be said that the notion of God does not solve any of the problems his theory of elementary particles is faced with. But surely, believing that a supernatural agency exists beyond our physical universe explains the eternal question of where this world came from? Besides this, assuming that our world has a purpose would help us anticipate a value for our existence and a real appreciation of the complex and sophisticated type of sense and consciousness that we are enjoying.

155 Altaie, M. B, and U. al-Ahmad (2011). "A Non-singular Universe with Vacuum Energy." *International Journal of Theoretical Physics* 50: 3521–8.

156 Weinberg, S. (1979). The First Three Minutes, New York: Bantam, 272.

SUMMARY OF THE PRINCIPLE OF TEMPORALITY

Temporality is associated with creation. It stipulates that the world and all its contents are created primarily out of nothing (*ex nihilo).* This is to say, that the world had a start *with* time. Eternity (*qadīm*) has no start with time. Temporality (*ḥuduth*) implies a continuity of creation, as originally envisioned by the Mutakallimūn.

The Mutakallimūn agreed that the whole universe was created, and, in this, they oppose those philosophers who say that the universe is eternal. The Mutakallimūn defended their views in this concern using a mostly rhetorical approach, although sometimes they supported their arguments with demonstrative ones. The demonstrative part of their theory of creation is best manifested in their theory of atomism, where we see the *jawhar* being associated with accidents, which are temporal and do not endure. This connection between the atomism of *kalām* and the principle of temporality, as well as other principles, is what makes *kalām* a whole consistent system of thought.

SCIENTIFIC SUPPORT FOR TEMPORALITY

Modern cosmology, which began in the 1920s with Edwin Hubble's discovery of the expansion of the universe, supports the theory that the universe could have come into existance at a particular moment of time (i.e. the Big Bang Theory). Later, it was found that the Theory of General Relativity, which was proposed by Albert Einstein in 1915, predicts this expansion of the universe. All the classical solutions of the Einstein field equations suggested models for the universe starting with singularity: a point with zero dimensions in space and time. However, when quantum effects are introduced, this singularity is avoided. Such quantum effects may include vacuum quantum fluctuations, out of which the whole universe would have been created in the presence of a curved spacetime background.

George Gamow and his collaborators investigated the natural abundance of ninety-two chemical elements in the universe, discovering that approximately 76% of it is hydrogen, 23% helium and only 1% is all the other elements. Gamow and his collaborators[157] tried to explain these ratios by suggesting that the universe about 14 billion years ago was in a soup-like state, which was composed of all the known elementary particles at a very high temperature. Particle and antiparticle pairs were created out of the thermal radiation at a very high rate but were soon annihilated to become radiation again, and so the story goes. As the universe went on expanding, its temperature dropped, and the more massive particles moved out of the thermal equilibrium. At first,

157 Alpher, R. A., Bethe, H., and Gamow, G., "The origin of chemical elements." *Physical Review*, 73(7), 803, (1948).

the creation of particle pairs stopped while annihilation continued. Due to the small difference in the mass of protons and neutrons, the latter decayed into protons, electrons and antineutrinos. Such a possibility increased the number of protons in comparison to the number of neutrons. Consequently, after about 3 minutes from the start of this scenario the temperature dropped to a level where atomic nuclei could form. At this time, the Gamow scenario suggests that the helium nuclei and the hydrogen nuclei, along with their isotopes, were formed.[158]

As the universe expanded, its temperature dropped further, and after about three hundred and eighty thousand years from t=0, the temperature became low enough to allow for electrons to combine with those nuclei and, thereby, form the first atoms. During this process of recombination, as it is called, radiation, due to the binding energy of the electrons, was released. This event caused the opaque universe to turn into a transparent medium, and once released, that radiation was transmitted through the universe without obstruction. This scenario predicts that the universe was overwhelmed with that radiation in the form of sourceless microwaves, which have a black body temperature of about 3 Kelvin, that is -270°C.

This theory of Gamow and his collaborators was able to show that hydrogen is the dominant element, at 76%, followed by helium, at 23%. The rest of the elements were found to be synthesised inside stars; a discovery made by Fred Hoyle and his collaborator.[159] Fred Hoyle disagreed with the Gamow scenario, calling it the Big Bang Theory.

An important point here, is that these background radiations, which were predicted by the Big Bang Theory, were discovered by Penzias and Wilson, who were surveying the sky for microwave sources.[160] A few decades later, in 1965, it was found that this background radiation contained a wealth of information about the very early stages of the universe. Current analysis of the spectrum of this radiation enables us to calculate the age, geometry, expansion speed, and matter energy content of the universe. The discovery of the cosmic microwave background radiation, as it is called, confirmed the hot origin of the universe.[161]

158 An excellent presentation of this scenario can be found in the magnificent book of Steven Weinberg, *The First Three Minutes* cited before.

159 Hoyle, F. Burbidge, E. Margaret, et al. (1957) "Synthesis of the elements in stars." *Reviews of modern physics* 29.4: 547.

160 Penzias, Arno A., and Robert Woodrow Wilson. "A measurement of excess antenna temperature at 4080 Mz." *The Astrophysical Journal* 142 (1965): 419–421.

161 Dicke, Robert H., et al. "Cosmic Black-Body Radiation." *The Astrophysical Journal* 142 (1965): 414–419.

Despite these, there are still several unanswered questions concerning the rate of the expansion of the universe and its matter energy content. It seems now that observable matter is only about 4% of the content of the universe, as predicted by its geometrical construct, which has been shown to be flat. As a consequence of this, cosmologists coined the term 'dark matter' to justify the missing mass of 96%. However, when it was discovered that the universe is expanding with acceleration, they invented another term called dark energy, splitting the 96% into 72% for dark energy and 24% for dark matter. In fact, these figures are modified periodically, depending on the data obtained from the different observations.

The established fact is that the universe has a beginning with space and time, although cosmologists are not absolutely sure whether the universe had only one start or whether it is cyclic. Einstein considered it to be a cyclic universe. If this is so, then it should go through a phase of collapse into a Big Crunch. In fact, the detailed picture of the creation and development of the universe is far from being understood.

The Cosmic Inflation

This is a theory which was originally proposed in 1982 by Alan Guth in order to remedy some fundamental problems of the Big Bang scenario. The anisotropy of the cosmic microwave background radiation suggests that the universe is flat, and has been flat since the early stages of its development. Inflation suggests that such flatness was acquired during the very rapid expansion that took place at the first stages of the universe, even before the creation of the first elementary particle. The theory also solves other problems that embodied the standard Big Bang Theory, such as the horizon problem, the flatness problem, and the magnetic monopole problem. However, despite more than thirty years having passed since this theory was proposed, cosmologists are uncertain whether there is any solid evidence in favour of the inflation theory.

I do not want to go into technical details here, but I should mention that the theory of modern cosmology is in need of serious revision. Among the major problems which need revision is the question of cosmic singularity, and the assumption that all the matter and energy content of the universe was created at a specific point in time. This is one of the main unsubstantiated assumptions of the standard Big Bang paradigm which is in need of revision.

THE NON-SINGULAR QUANTUM MODEL

As mentioned above, an alternative model which takes into consideration quantum effects, the creation of the universe and its development at the very early stages, proposes a growing universe that started as a single, incredibly tiny point with a critical density, and expanded while its energy matter content was growing in proportion to its size. The model is based on the idea that a sudden curvature was made available within an otherwise empty space time. This sudden curvature promoted the conversion of virtual particle-antiparticle pairs into real positive and negative energy states. The negative energy was absorbed by the positive curvature of spacetime, causing it to expand and become flattened, while the positive energy was converted into particle-antiparticle pairs. A scenario similar to that of Gamow and his collaborators then began. The model shows that a universe with a critical density was created out of this process. In this model, the density of the universe decreased at a rate proportional to the square of the radius of the universe. The model is free from the problems of the standard model, including singularity; which disappears because of the quantum effects that accompanied the birth of the universe.[162] This model has not been taken into consideration seriously enough by theoreticians, even though it might fit better with the observations; but, unfortunately, it is the prevailing dogma that is dominant.

In any case, the predominant facts about our universe confirm that it is temporal in every respect, as well as in the formation of all of its constituents.

162 Altaie, Mohammed Basil, and U. al-Ahmad. Loc. Cit.

Chapter Five

The Principle of Re-Creation

This is one of the most important principles of *Daqīq al-kalām*. It originates from the postulate that the transients (*aʿrāḍ*) do not endure in two instances. This assumption was adopted by most of the Mutakallimūn, especially the Ashʿaris, but it is said that Muʿtazilis were divided about this question, with some reported to have accepted the immutability of some of the transients and suggesting the non-endurance of other transients.

Maimonides explains the assumption of non-endurance of the transients by saying that, "once an accident is created it goes away and does not stay, then God creates another transient of the same type and this goes too and a third one of the same type is created, and so on as long as God wants that transient to stay", then he says, "If God wants to create another type of transient in that *jawhar* He will do, but if He stops re-creating the transient and does not create a transient, the *jawhar* will cease to exist. This is the view of some of them [the Mutakallimūn] who are the majority, and this is the creation they talk about."[163]

Al-Ashʿari reports in *Maqālāt* that there are some of the Muʿtazilis, like al-Shatwī, al-Kaʿbī and al-Ṣayramī, who claimed that transients do not stay, but some others, like Abū al-Hudhayl al-ʿAllaf, who say that some of the transients may stay. Al-Ashʿari tells us that al-Jubbaī used to say that, "colours, flavours,

163 Maimonides, *Guide for the Perplexed*, 202.

smells, life, power, and health stays."[164] From the other side, Ibn Matawayh reports that Muʿtazilis hold that, "colours, flavours, smells, heat, coldness, moisture, dryness, life, power, the being (*kaūn*) and composition (*ta'līf*) stay, and whatever else there is of these types, should not stay."[165] Maimonides suggests that the motive behind holding this doctrine by the Mutakallimūn is to establish their view of negating the existence of active intrinsic nature, which causes events to happen in the world. He says, "The object of this proposition is to oppose the theory that there exists a natural force from which each body derives its peculiar properties. They prefer to assume that God himself creates these properties without the intervention of a natural force or of any other agency: a theory which implies that no transient can have any duration."[166]

Ibrahim al-Naẓẓām is reported to have argued against all the transients except motion.[167] Generally, the Muʿtazilis are known to hold that *some* of the transients may endure, meaning that they advocate the *jawhar* remaining unchanged.

The Egyptian researcher Muna Abū Zayd claims that such a trend among the Muʿtazilis was initiated by the requirement of having human free will, since if the transients are to be renewed, then it would be Allah who is renewing them. Consequently, Allah will be responsible for the acts of humans, which is in contradiction with their belief. This opinion is shared by Al-Iraqī who finds that the negation of the Muʿtazilis for the endurance of some accidents is due to the contradiction between human freewill and the re-creation of the transients and the *jawhar*.[168]

This explanation by al-Iraqī and Abū Zayd is arguable as they have not provided any substantial evidence from the Mutakallimūn regarding this question. On the other hand, the renewal of the transients need not contradict human freedom since it is the human himself who is deciding on what to do. For example, I may say to a person: *I will give you $1000 to invest in the market and it would be up to you to decide on whether to sell or buy, but I will renew the capital from time-to-time, so that it will be kept at a certain level.* In this case, it will be the decision and responsibility of the man for any gain or losses he might incur. Renewal of the capital acts to fuel the decisions of the person in charge of the investment but he does not share the responsibility of the decisions taken, which leads to either losses or gains. These remain the responsibility of

164 Al-Ashʿari, *Maqālāt al-Islāmiyyīn*, 359.

165 Ibn Matawayh, *al-Tadhkirah*, 41.

166 Maimonides, *Guide for the Perplexed*, 202.

167 Al-Ashʿari, *Maqālāt al-Islāmiyyīn*, 358.

168 Al-Iraqi, A. (1973). *al-Tajdīd fi al-Mazahib al-Kalāmiyyah wa al-Falsafiyyah*, Cairo: Dar al-Maʿarif, 70.

the decision maker. It would be legitimate, then, to hold the person responsible for his acts, with no blame whatsoever attached to the subsidising agency (myself), who is renewing the capital.

The only excuse for claiming that the re-creation of transients can influence human free will is through the re-creation of the will itself, and this might lead to negating the renewal of will, since it would certainly mean that freewill would be subject to the will of the Creator. For this reason, it might be worthwhile to re-visit the views of the Muʿtazilis on this issue in more detail, in order to scrutinise the claim behind negating the re-creation of some transients. However, it should be kept in mind that this question of renewal of the will, and all other non-physical entities, is not a subject of the new *Daqīq al-kalām*.

TRANSIENTS ACCORDING TO ASHʿARIS

Both the Ashʿaris and Māturīdīs believed in the re-creation of all transients; this was one reason why they were considered occasionalists. Al-Bazdāwī, subscribing to the Māturīdīs, affirms that the non-endurance of transients is the belief of "Sunni Muslims",[169] and al-Bāqillānī has expressed this Ashʿari view by saying that, "The accidents are those that do not endure and that emerge in jawāhir and bodies, and cease the next moment of their existence."[170] Therefore, we should recognise that the transients are emergent properties associated with the jawāhir and the bodies.

For the temporality of the transients, al-Bāqillānī argues that, "the evidence for proving the accidents is the movement of the body after being at rest and then its returning to rest after it was in motion. This could be either caused intrinsically or for a reason. Should it be an intrinsic property, the body would never be at rest. Accordingly, the body should have been caused to move."[171] In fact, the emergence of transients as being re-created properties reinforces the view of the Mutakallimūn that the world is a renewable entity, whose existence is due to this state of being under continuous re-creation. On the other hand, the assumption of the re-creation of transients is closely related to the doctrine of the temporality of whole world. In this respect, Wolfson recognises that the argument of al-Bāqillānī "consists of a number of successive propositions, each of which is established by proof."[172] Starting with the proposition that: "the world, the upper and the lower, is inseparable from two genera, namely,

169 Al-Bazdawī, *Uṣul al-Dīn* edited by Hans Peterlis, Cairo: Dar Iḥya' al-Kutub al-ʿarabiyah, 12.

170 Al-Bāqillānī, *Tamhīd*, 38.

171 Ibid., this example might seem inadequate since such a description applies to bodies not the *jawahir*.

172 Wolfson, *kalām*, 393.

jawāhir and transients,'[173] Al-Bāqillānī proceeds to prove its creation by providing three propositions:

1. that transients exist in jawāhir and bodies
2. that transients are temporal properties (*ḥawādith*), and
3. that bodies are created, which is proved by the reasoning that, "they do not proceed after their temporal properties (the transients in them) and do not exist before them," and, according to al-Bāqillānī and the rest of the Ashʿaris, "whatever does not precede that which is created, is created like it." This shows how closely and intimately the principle of the re-creation of transients is related to the principle of temporality of the world.

Analysis of the Principle of Re-creation of Transients

Here I would like to expose the theological as well as the scientific value of the principle of re-creation which, as I have shown, was adopted by both the Ashʿaris and the Muʿtazilis. In fact, the principle resolves several theological and philosophical problems. These are:

1. The realisation of the role of the Divine in the world, with the process of re-creation asserting the role of the Creator in sustaining the universe.
2. The realisation of natural phenomena (laws of nature) as a probabilistic character, rather than being a deterministic set of orders that are firmly applied independent of any other agency.
3. It solves the problem of the relation between the Creator and his creation by maintaining their connection through the process of re-creation.

From the natural philosophy side, the principle of re-creation provides us with a picture of a dynamical, non-static and mutable world that is under continuous change; a property which makes it, at every moment, turn anew. The implications of the notion of re-creation are:

1. That the physical values for quantities that we observe in nature are under continuous change and it might be that what we actually measure is an average of all possible values.
2. That the world is open to contingencies that would be available for the re-created objects. This, however, needs the additional postulation which we see adopted by the Mutakallimūn.
3. It would also imply that some sort of uncertainty is embedded in the natural world due to changes occurring in the values of some observables. This is not self-evident but will be demonstrated in more detail later.

These implications are not self-explanatory results but are such that one should be able to deduce them from the principle of re-creation. It is akin to theorems that can be proved once an axiomatic system is set with clear suppositions.

173 Al-Bāqillānī, *Tamhīd*, 41.

For this reason, I will present these results in a deductive approach and take them as an implication of the re-creation principle. Theologically, the principle of re-creation assumes the continuity of Divine action In the world. The Mutakallimūn is philosophy on this has been presented elsewhere in this book, and there is no need to repeat their argument here.

LAWS OF NATURE

In traditional *kalām*, the laws of nature are understood to be regularities that we are accustomed to seeing occurring. This is what makes a phenomenon, and will be discussed under the topic of causality in Part Three of this book. However, since the laws of nature are not identified clearly in terms of their specific character in *Kalām*, I feel the need to discuss it here.

In our everyday world, we see phenomena occurring naturally. Through careful observations, follow up and measurement, man has been able to understand the causes behind some of these phenomena and to identify the relationships between the involved variables which allow them to occur. For example, cotton burns when approached with fire and gains enough heat to reach its ignition temperature. Once ignited, cotton will go on burning unless it has no access to air (oxygen). This is what I call a *law of nature*, as it is expressing a well-defined natural regularity that takes place whenever all the required conditions are available.

Another example is the force of gravity and freefall. If we throw a stone upwards, it will ascend with decreasing speed until it reaches a maximum height, at which instance, the velocity of the stone becomes zero. For a moment, then, the stone is at rest, before it falls back towards the ground, this time moving with increasing speed which reaches maximum value when the stone hits the ground.

Galileo Galilei studied this motion quantitatively, leading to the deduction of the laws of freefall and identifying the acceleration by which the stone descends toward the ground. He found, contrary to what Aristotle believed, that this acceleration was independent of the mass of the stone and has the same value for all freely falling bodies. The Mutakallimūn discussed this motion in qualitative details.

Galileo also analysed the motion of a simple pendulum, composed of a metal bulb suspended by a length of string and found that this motion obeys the same laws as a freely falling body. During the second half of the 17th century, Isaac Newton, studying the work of Galileo and others, was able to symbolise the law of free fall and other phenomena studied by Galileo. The result was his three laws of motion and the law of universal gravitation, by which Newton was able to understand why planets revolve around the Sun in elliptic orbits, and explain the three empirical laws which were deduced by Johannes Kepler after a lengthy analysis of the astronomical observations of Tycho Brahe.

This very brief presentation of some of the fundamental phenomena in nature tells us that natural phenomena are regular events which take place according to a well defined set of rules. We call these rules the laws of nature. These laws, as it seems, occur on a deterministic basis; once the conditions are made available for a given phenomenon, then it is certain to happen. It was during this period, from the 17th century onwards, when scientists began to comprehend the world using proper means and the correct approach, that modern science was born.

THE ACTION OF THE LAWS OF NATURE

It was this discovery, by which I mean the methodology of modern science, which made people comprehend that what controls the world is not those absent Gods, but the laws of nature with their active spontaneity and clarity.

Using the advanced approach of this modern natural philosophy, further developments along this line of thought occurred during the 18th and 19th centuries. These developments provided mankind with so much precise knowledge about the world that people thought the universe to be a gigantic machine run by the laws of nature which are independent of any supernatural power or intervention. This enabled Pierre Laplace to set his understanding of the world as being comprehensible in its past, present and future. He says:

> We ought to regard the present state of the universe as the effect of its antecedent state and as the cause of the state that is to follow. An intelligence knowing *all* the forces acting in nature at a given instant, as well as the momentary positions of *all* things in the universe, would be able to comprehend, in one single formula, the motions of the largest bodies as well as the lightest atoms in the world, provided that its intellect were sufficiently powerful to subject *all* data to analysis; to it nothing would be uncertain, the future as well as the past would be present to its eyes. The perfection that the human mind has been able to give to astronomy affords but a feeble outline of such intelligence.[174]

This statement of Laplace's reflects the degree of confidence in the laws of nature and their deterministic character. The laws of nature are qualitatively understood when we formulate them using the symbolic language of mathematic. This language expresses the relationship between the physical variables that we recognise related to the phenomenon it is describing, and this is what we call the *laws of physics*. Hence, the laws of physics are actually our mental

174 Laplace, Pierre. *A Philosophical Essay on Probabilities*. F.W. Truscott and F.L. Emory (Trans.). New York: Dover 1951.

construct for the laws of nature. These laws are not fixed but may change, depending on our comprehension of the laws of nature. This is a very important point, since there is much confusion about it in literature which tends to focus on popularising and simplifying science.

The unfounded assumption that the laws of nature are capable of acting independently and affecting causes for events to occur in nature, allows for deterministic causality and the denial of any need for an external intervention beyond the physical system and its variables. The world is thought to be a stand-alone block universe that does not need a Creator or Sustainer. Such a conclusion is consolidated with the notion that the universe could have existed endlessly in the past without having a beginning. This is what Stephen Hawking questioned when he discovered that the universe might have no beginning but could have existed for an endless, imaginary time. Hence, his rhetorical question, "what place, then, for a Creator?"

Such a question misses the fact that the laws of nature do not act on their own since they have no actual self-effectiveness. There is no evidence or rigorous proof that confirms that the laws of nature act on their own. This is an age-old question that has been discussed by philosophers such as al-Ghazālī and Hume, and culminates with the metaphysical understanding of Nancy Cartwright.[175] The argument is consolidated by the fact that, according to quantum physics, natural phenomena are not deterministic events; they are probabilistic. Therefore, the laws of nature cannot be considered, reliably, to play an active or intrinsic role since they lack determinacy. On the other hand, the laws of physics are a mere description of the laws of nature; rather it is our own mental comprehension which is fallible. Newton's description of gravity has been shown to be completely inaccurate on both the conceptual and mathematical levels, and has been replaced by Einstein's general relativity description.

The other problems that the re-creation principle solves is the question of the relationship between God and the world, as well as the age-old question that stems from the dilemma of having a metaphysical entity contacting and influencing physical quantity. Indeed, how can we imagine such an effect to take place unless there is the means by which the physical is actually influenced by the non-physical? The fact that matter and energy in the universe are known to be conserved suggests that no influence can be imagined to take place unless it comes from within the universe. Such questions have been raised by scholars such as Keith Ward, Professor of Divinity at Oxford University, and John Polkinghorne, an ordained priest and the ex-Professor

175 Cartwright, Nancy. "No God; No Laws." In Dio, la Natura e la Legge: God and the Laws of Nature, edited by E. Sindoni and S. Moriggi, 183–90. Milan: Angelicum-Mondo X. (2005).

of Particle Physics at Cambridge University. Ward believes that God does not interfere with the world in the details, that He may not even know the future, and may not be aware of the suffering of people.[176] Polkinghorne, on the other hand, asserts that God may intervene in the world by His word or His order, maintaining changes through the energy available in the world itself. God, according to Polkinghorne, issues his orders as an epistemology—a sort of passing on information without the need for energy—that is addressed to provide a model for ontology. The energy needed for the ontological changes is utilised from within the universe itself.[177] While this is an intriguing idea, it is in need of a detailed mechanism to achieve it. The interface between the order of God and the physical world is still not clear.

RE-CREATION: WHAT DOES IT EXPLAIN?

Re-creation of the transients is a possible and profound answer to these questions, including the establishment of the relation between the Creator and the created, both instantaneously and continually. As the Creator is renewing his creation every moment, he must be keeping in touch with it. As such, the world is maintained temporally while the Creator is eternal. It would be interesting to point to four basic characters of our physical world that re-creation can explain. These are:

1. The principle of re-creation explains, for us, the operational construct of quantum mechanics. This is brought about by the simple fact that re-creation is an operational process by itself, in which several operators may be at work, and since all physical quantities that we are measuring are the re-created transients themselves—as described by the *kalām* expression—the process of the measurement, therefore, is only related to the measured object itself. This is what, in physics, is called the 'observable.'
2. The process of re-creation requires that things are being made anew every moment. This allows for claims that the values of the physical observables are under continuous change, thereby forming a set of possible values. The size of such a set of values depends on the system under consideration, and could be finite in number or infinite. This explains why we observe a spectrum for the so-called eigenvalues for any observable in quantum physics. Normally, in standard quantum calculations, we obtain such eigenvalues by applying a mathematical operator on what is called the wave function (or state vector in the language of matrix mechanics).

176 Ward, K. Pascal Fire: *Scientific Faith and Religious Understanding, Oxford*: One World, (2006).

177 Polkinghorne, John. "God's Action in the World." J.K. Russell Fellowship Lecture, Pacific School of Religion Chapel, Berkeley, CA, 6 April 1990.

3. What we measure every time is not these discrete eigenvalues; otherwise, we would not be able to find the system in one and the same state. For example, when we measure the energy released by, say an atom in a given state, we always find the same value. This reflects that we are measuring an average value obtained out of a set of values, each of which might not be accessible individually for the measurement. The average value in quantum mechanics is called the *expectation value*. The higher the energy of a system, the nearer is the measured value from the expectation value. This explains why physical quantities in the macroscopic world are much more determined than those which we find in the microscopic world. The fact that the re-creation rate of observables in the macroscopic world is very high is exhibited by the higher determinism level of this world.
4. This leads us to expect that there must be some uncertainty in quantum measurements, as re-creation would mean creating a new state every time. Such uncertainty is expected to occur upon measuring observables belonging to *complementary operators,* such as position and momentum, energy and time, angular momentum and angular position, etc. This is expected to be the case for all observables that are generators of each other. So far, any pair of observables that are known to be generators of each other would be expected to have some inherent uncertainty upon the simultaneous measurement of the two observables. This means that no such pair of observables can be measured simultaneously with infinite accuracy. This might explain the origin of the Heisenberg uncertainty principle. However, here a question arises about the body which is at rest, that is: would the re-creation of accident create some kind of motion in it? The answer is yes, indeed. Inevitably, one would expect that as the value of the position (and momentum) changes upon re-creation, then a degree of local motion would occur. This can be understood by imagining that the re-creation of the position of the body will allow it to occupy a new position on the right or the left, to the front or the back, above or below, with respect to the original position of the body. All these are equal possibilities. Thus, if re-creation is taking place, then the state of absolute rest cannot be maintained. Indeed, this is what quantum mechanics predicts; that the expectation value of the lowest kinetic energy of a body at rest in a bound system is non-zero. This reminds us of what al-Naẓẓām said: "I don't know what rest is, except that the body was in that position at two separate instances, meaning that it moved through it two times."[178] Perhaps the explanation given above would support Max Jammer's claim that, "In fact, al-Naẓẓām's notion of leap, his designation of an unanalyzable inter-phenomenon, may be regarded as an early forerunner of Bohr's conception of quantum jumps."[179]

178 Al-Ashʿari, *Maqālāt*, 325.

179 Jammer, M. *The Philosophy of Quantum Mechanics: The Interpretations of Quantum*

The above explanations might contribute to clarifying the basic characters of the quantum world and enable us to comprehend why the quantum world is so fuzzy and disturbing, as to make us wonder about its objectivity and its reality.

INTERPRETATIONS OF QUANTUM MECHANICS

I have tried to utilise the principle of re-creation to suggest a resolution to the problem of measurement in quantum mechanics. This is one of the most difficult problems in quantum physics. In some respects, the problem is philosophical rather than technical, as it is directly related to the measurement process. The different interpretations have added more confusion to the problem and caused a misunderstanding of the original concepts involved. This problem has been presented with excellent background details by Max Jammer in his afore-mentioned excellent book, *The Philosophy of Quantum Mechanics*. The problem has many facets, the most important of which are:

1. The probabilistic distribution of the measured values, and their mutability, which characterise the indeterminism of the measured results (occurrence of events).
2. The inherent uncertainty in the measured values of the so-called incompatible observables.
3. The non-commutativity of the mathematical operators corresponding to the incompatible observables.

One should mention that the most important aspect of all this is the inherent indeterminism which makes the occurrence of events in nature somewhat uncertain and shows that it is as if God plays dice; a position which Albert Einstein strongly rejected.

There are many interpretations that were suggested for the process of quantum measurement, each of which reflects a different understanding of the theory and the employed techniques. One interpretation says that quantum mechanics can be understood as the mechanics of quantum ensembles, not that of individual particles. Einstein and Max Born were perhaps the strongest advocates of this opinion.

An alternative suggestion is that a quantum measurement is one taken when the state of the system collapses in one of the possible states allowed by the physical solutions available for the system. John von Neumann provided the mathematical formulation for this explanation, which was called the wave function collapse. This interpretation involves the interaction between the observable and the observer, and requires the participation of the observer in modelling the outcome.

Mechanics in Historical Perspective. New York: Wiley-Interscience,, 259 (1974).

A third interpretation suggests that there are some hidden variables which play a role in determining the result of quantum measurements. Such variables are complementary to the quantum calculation and are important for determining the result, which is thought to be deterministic. The scientist who first proposed this was David Bohm.[180] He believed that the standard formulation of quantum mechanics was incomplete and should be amended by the assumption of hidden variables.

A fourth interpretation was suggested by Hugh Everett[181] in the early 1950s. He proposed that physical states, which are being measured, split spontaneously into an infinite number of states, each of which belongs to a world that exists in parallel, and simultaneously, with our own physical world, and that, upon making the measurement it just so happens that we just pick up one of those possible values by being in one of those worlds. This interpretation gave rise to the hypothesis of multiverse (or many worlds).

Each of the above four interpretations has a weak point. The first one is refuted by the fact that quantum calculations, as applied to single particle systems, work very well and are compatible with a number of experiments. The double slit interference experiment, for example, performed by sending one electron at a time, provides verification that quantum mechanics applies to single particles theoretically as well as experimentally.

The Copenhagen interpretation, devised by Bohr and Heisenberg, may have once seemed robust but has since had objections levelled at it; it does not suggest any mechanism other than the instantaneous collapse of the wave function, such that the reality of the state can be observed at the moment of the measurement. It is not clear how this collapse happens or why, and it might be absurd to accept the role of the observer, who, it is claimed, can determine the results of the measurement. Such a role has been exemplified by John Wheeler's claim that later observers may affect the results which occurred earlier. This explanation, i.e. the collapse of the wave function and the role of the observer in determining the result of a measurement has been severely criticised by Erwin Schrödinger through the well-known Schrödinger's cat paradox.

Bohm's interpretation, involving hidden variables, tends to attack the indeterminism of quantum mechanics by trying to replace it with a deterministic theory. This interpretation received a blow when work, conducted by Alain

180 Bohm, D "A Suggested Interpretation of the Quantum Theory in Terms of 'Hidden' Variables, I and II." *Physical Review* 84 166–79. (1952).

181 Everett III, Hugh "Relative State Formulation of Quantum Mechanics." *Reviews of Modern Physics* 29: 454–62 (1957).

Aspect, verified Bell's theorem and found inconsistencies with the local hidden variable theories. On the other hand, Leggett's theorem, which refutes the non-local hidden variables, was verified experimentally.

Everett's proposal might be more acceptable than the others, as it has the loophole of realising many worlds which exist simultaneously infused within one another. However, this is quite absurd, and is a big challenge to reality and objectivity.

Chapter Six

Contingency and Indeterminism

The principle of contingency is one of the basic doctrines of *Daqīq al-kalām*, and one on which the Mutakallimūn's view of the physical world rests. The principle is intimately related to the question of causality and which, along with the previous three principles, atomism, temporality and re-creation provide a consistent understanding of phenomena and events taking place in the natural world.

In this chapter, fundamental statements concerning this principle will be given, analysed and discussed. The main points are that, contrary to common belief about the diverging views of the Ashʿaris and the Muʿtazilis about the question of natural causality, we find that in-depth analysis, which is supported by the views of their exponents and found in authentic sources, reflects a broad agreement on the basic premise of understanding natural causality.

Both the Ashʿaris and the Muʿtazilis acknowledge causal relationships and negate causal determinism. The main difference between the two schools is on the question of human causality, where the Muʿtazilis hold that man is a causal agent who is deemed responsible for his deeds, while the Ashʿaris hold that God is the sole Creator of everything, including the actions of man. They believed that man was responsible for what was gained by his actions but that he was not the creator of any of them. This was the theory of *kasb,* which was defended by the Ashʿaris. As for causal relationships, the Muʿtazilis are known to have elaborated on these, describing them in two more terms in addition to the one held by the Ashʿaris, which is *ʿādah* (custom). Consequently, there are some differences that lie in explaining the mechanisms by which causal relationships work.

The most fundamental concepts involved in this principle include (but are not limited to) the following: nature, necessity, cause, effect, determinism and contingency. The theoretical foundation for the content of this principle is based upon:

1. Assuming the re-creation of accidents, and
2. Negating the existence of an intrinsic effective nature within the inanimate matter.

Al-Shahrustānī says, "the mental quotient restricted knowledge into three categories: the necessary, the impossible, and the contingent. The necessary must exist because if it did not it would cause an impossible, and the impossible is necessary nowhere; so if it is estimated to exist, then it would cause an impossible, and this contingency is neither necessary nor impossible."[182] Prior to this, he said, "If it is asked, 'what is the evidence on which the existence of contingent would have touched the whole world?' In reply, we say that the frank mind requires contingency in each one of the parts of the world, and since the total is composed of the parts, contingency would be then imperative."[183] Accordingly, I may add that the principle of contingency in simple words is that, "Every part and every event in the world is deemed contingent unless it is impossible."

The theological basis for this principle is the belief in the role of God as Sustainer of the world. The argument put forward by the Mutakallimūn for adopting this principle is based on:

1. the notion that inanimate matter is impotent and has no power or will, which is needed for any self-action.
2. that the conjunction between cause and effect is indecisive on the necessity of these relations, which makes the effect indeterminate.

We will see that most of the Mutakallimūn agreed on the main views concerning this principle, but they differed on some details. For example, the Muʿtazilis invented something they called *tawlīd* (generation) in order to explain the attribution of an effect that is caused by some secondary agent or

182 Al-Shahrastānī, Muḥammad ibn ʿAbd al-Karīm *Nihāyat al-Iqdām fī ʾilm al-kalām*, Alfred Jeyom (Trns.), Oxford (1934).
183 Ibid.

activity. Also, they devised the term *iʿtimād* (adherence) in order to explain some causes that were called effects of concerning forces, such as the gravitational effect on a stone thrown upwards. These detailed differences might have caused some conflict between the Muʿtazilis and the Ashʿaris. However, as far as their basic belief concerning the sole role of God in sustaining the universe was concerned, they were in full agreement.

The *ṭabʿ* (Nature)

The question of intrinsic nature is a pivotal concept in dealing with the problem of contingency and necessity in *kalām*. In what follows, I will present the views of traditional Mutakallimūn on this issue starting with prominent Muʿtazilis and their elders.

Definition of *ṭabʿ*

In the Arabic lexicon, *ṭabʿ* is defined as, "the nature or genus according to which man is patroned."[184] Al-Jurjānī says, "*ṭabʿ* is the genus with which man was created, and nature is the force which runs through the body by which the body achieves its natural perfection."[185]

Nature is a derivative of the term *ṭabʿ*, and the genus of man is his behaviour, while his nature includes more basic aspects, e.g., his shape, the colour of his skin and other such attributes. The *ṭabʿ* is meant to be the mould of man from which he is cast, so to speak. Also, it is noted that the term *nature* refers to all that is in the world other than man, and I gather that this term is meant to indicate an intrinsic property that the thing is cast upon, like saying that the nature of fire is burning, and the nature of gravity is attraction. So, therefore, nature is quite a suitable interpretation for the word *ṭabʿ*.

THE VIEWS OF THE MUʿTAZILIS

Some authors have confused views of the Muʿtazilis in regard to human freedom and their attitude towards the responsibility of humans for their acts. Consequently, people thought that the Muʿtazilis believed in the existence of some intrinsic nature in things, in the determinism of the world and the deterministic causality. All of this is inaccurate, as we will see.

We have some original sources of the Muʿtazilis' legacy made available during the last few decades, through which we gain an accurate picture about their attitude towards the question of intrinsic nature and their understanding of determinism of the world. Obviously, it should be emphasised that here I mean the majority of the Muʿtazilis, and not necessarily all of them.

---

184 Al-Rāzī, Zayn al-Dīn, Mukhtar al-Sihaḥ, edited by Yousef Shaykh Muḥammad, Beirut: al-ʿAṣriyah printing shop, 1999.
185 Al-Jurjānī, al-Tarīfāt: al-Tab.

Some of the main, authentic sources are the books of Qaḍī ʿAbdul Jabbār, notably, *Mughni fi abwāb al-Adl wa al-Tawḥīd* and *Kitab Al-Mahit Bil Taklif.* Also, we have the book of Ibn Matawayh, *Al-Tadhkira fi ahkam al-jawāher wal arʿād.* The book of Abū Rashid al-Naysābūrī, *Masāʾil,* stands as a very good source for understanding the detailed differences between the Basrian and Baghdādī Muʿtazilis.

In addition to this, there are the sources of Ashʿaris which reported some of the views and discourse of prominent Muʿtazilis, such as al-Naẓẓām and Muʿammar. The former affirmed the existence of secondary causes while confirming the role of the Divine in maintaining and sustaining the world. In his book, *al-Intiṣār,* al-Khayyāṭ reports that al-Naẓẓām said: "If it is in for the water to flow, it is in the ability of God to prevent it from doing so, and if it is for the heavy stone to fall then it is in the ability of God to prevent it from that."[186] Primarily, this shows that, despite all that has been said about al-Naẓẓām with regards to his belief in the determinism of nature and deterministic causality, the fact is that he only deviated from mainstream Muʿtazilis' opinion on *Daqīq* al-*kalām* in a very few points, such as the suggestion of *ṭafra,* for example.

However, it seems that Muʿammar was a true naturalist, as we see when ʿAbdul Jabbār argues against his approach of attributing events to the nature of things. He says, "perhaps the Dahrīya (eternalist) can be excused in comparison to naturalists like Muʿammar and his followers, because those (the Dahrīya), when they negated the Creator, requested an alternative to which events might be attributed. But what excuse can be given to those who affirm the existence of a free Creator while failing to attribute these [events] to him and instead attributing it to something irrational, despite knowing how to relate the act with the powerful."[187] Here, we see that ʿAbdul Jabbār does not attribute the occurrence of an event to any natural property processed by things, and is blaming Muʿammar for being a believer while not attributing events to an able Creator. Furthermore, ʿAbdul Jabbār has made it clear that the Muʿtazilis do not endorse the claimed actions of intrinsic nature. He says, "And when they [the philosophers] say of 'Nature' that which happens from the burning fire, it is what we consider happening out of the adherences (*iʿtimādāt*), which generate dissociations. This is as if they call Nature what we call *iʿtimād.* This is similar to what happens to weight when immersing a heavy [body] in water, generating a fall, and other such things. We attribute all these [events] to an actor who is free and able to prevent such generation and response."[188]

186 Al-Khayyāt, Intiṣār, 48.

187 ʿAbdul Jabbār, *Muḥit*, 102.

188 Ibid. 101.

VIEWS OF AL-BĀQILLĀNĪ

In his book, *Tamhīd*, al-Bāqillānī studied the concept of *ṭab*ʿ (nature) and the possibility of its action on affecting events. He discussed the issue from three positions:

1. in the first, he denies that the Creator of the world could be any form of nature that requires the existence of the world.
2. negating that the world is composed of the four natures, which are heat, cold, moist and dryness.
3. refuting that the Creator of the world and its Sustainer to be the seven orbs, as claimed by astrologers.

In what follows, I will present the views and arguments of al-Bāqillānī, through which we may understand his position on causality and the occurrence of events in the world.

The first problem: the Creator of the world

The argument here is centred on disproving that the Creator of the world could be any form of nature that necessitates the existence of the world. This is presented as a rational approach, using the standard methodology of argument in *kalām*. The question posed is whether this alleged nature is eternal or temporal. Al-Bāqillānī explains that if it is to be temporal, then it cannot create things since the temporal is unable of creating, "If it would be eternal, then the created events should be eternal too."[189]

This might seem puzzling since the Mutakallimūn, and al-Bāqillānī, in particular, believed that God, the Creator, was eternal too. So, how can it be that the event cannot be created out of an eternal nature unless these created events are eternal too? Here, we need to clarify how the Mutakallimūn, in general, differentiate between the willing, omnipotent, and omniscient Creator and the inanimate natures which might be claimed to have the ability for effecting a change or creation. This paradox is resolved once we know that the Mutakallimūn believed that whenever an inanimate object has a nature to act, then that nature should be acting all the time since inanimate objects do not have the will required to choose when and why to act. Accordingly, such an understanding was built within *kalām* logic, which assumes the necessity for will in order to be able to act. If this will is available, then action can be taken, and can also be stopped. If the will is not available, then action is impossible. If it happens once, then it should (say according to some natural ability) act in the same way forever. Logically therefore, it is the presence of will that controls actions. As inanimate objects do not have a will, they should not be able to act creatively. Some other entity which has both will and ability should act.

189 Al-Bāqillānī, *Tamhīd*, 53–4.

Adopting this, the Mutakallimūn were able to discuss and refute several arguments of the philosophers. In his discussion, however, al-Bāqillānī criticises the attitude of the Muʿtazilis for adopting *iʿtimād* (adherence). This is a sort of secondary ineffective agency that is proposed by the Muʿtazilis in order to explain causal relationships. Al-Bāqillānī thought this might imply that an effective agency was acting intrinsically in the world. Al-Bāqillānī and the other Ashʿaris rejected the existence of *iʿtimād* (adherence) since this would mean the existence of some kind of intrinsic effective nature, which directly contradicts the Islamic creed.

In my opinion, this is not a good stance to take as it had already been shown, by the Muʿtazilis, that their belief in God, the Creator, did not contradict, assuming the existence of such things as *iʿtimād* and *tawlīd*. However, the logical necessity for adopting such concepts was the need to establish causal relationships. Besides this, as we have seen earlier, ʿAbdul Jabbār made it clear that he also used the term custom (*ʿādah*) to refer to the regularities in nature, without associating it with the *iʿtimād* or *tawlīd*. In the new *kalām*, this issue receives the proper treatment, and gets its proper role in clarifying causal relationships. The approval of the existence of causal relationships does not contradict the basic belief that Allah is the sole Creator of events and the active Sustainer of the world, but, certainly in the new *kalām*, Allah is not understood as a tyrant puppeteer. However, when it comes to discussing human causality, the issue takes on a different meaning and content since man enjoys certain abilities, along with a will that has been obtained by virtue of his creation. This entitles him to be a secondary Creator, so to speak. This is, perhaps, what the Qur'an inferred when stating in Surah Al-Mu'minūn, "*So blessed be Allah, the Best of Creators!*" (23:14).

The second problem: the four natures and elements

Al-Bāqillānī rejected the notion that the world is composed of four elements, or natures, which the Greek and some Muslim philosophers, such as Ibn Sīnā and Ibn Rushd, had adopted. He sets out his arguments through a series of aspects.

The first of these aspects is to consider the four natures as accidents, and since the accidents, according to the Mutakallimūn, do not act, al-Bāqillānī finds those natures impotent and maintains that they cannot constitute active elements in the composition of the world. The second argument that al-Bāqillānī presents uses his previous demonstration, which shows that those natures are temporal, saying that if the world is to be composed of such elements, then it cannot be considered eternal. In the third argument by which he claims to refute the effectiveness of those natures, al-Bāqillānī resorts to *kalām* atomism, utilising the property of the jawāhir of being identical. This being so, he then says, "as such, all effects due to such jawāhir should be similar." Consequently, the effects due to one object should be identical to the effect produced by any other object or entity. As an example, he continues, "if it was the same material [causing the state of being drunk] then intaking any other stuff will cause the same effects as drinking… and the neighbouring body [to another hot or cold body] should result in cooling and heating it, since we have demonstrated that all bodies have the same composition."[190]

Al-Bāqillānī asserts the view of the Mutakallimūn that real acts can only emanate from an omniscient and omnipotent God, and not from inanimate matter, because these are composed of jawāhir and accidents. He further states that the jawāhir cannot affect acts as they are of one type while the accidents cannot act on their own. It is important to recognise how the Mutakallimūn sterilise the natural properties of things from their claimed effectivity using their atomism.

The third problem: astrology

It was a common belief amongst astrologers and most Greek philosophers that celestial objects exert a great influence on, what they called, the lower world below the orb of the Moon. Muslim philosophers, including al-Kindī, Ibn Sīnā, Ibn Rushd and al-Fakhr al-Rāzī, believed in this description, in one way or another. The Mutakallimūn argued against such a claim when they dismissed the artificial division of the world into an upper and lower part. Al-Bāqillānī vigorously denied that the astronomical orbs could be the caretaker of the lower world, "because they follow the same formation of the other bodies in the world, which have finite limits, composition, motion, rest, and the transformation from one state into another; as happens to all other bodies in the world. So should it be possible to have been eternal, then all other bodies would have been eternal too."[191]

The most important part of this quote is the belief of al-Bāqillānī and the rest of the Mutakallimūn that celestial objects are similar to all other objects in the world; meaning that they can corrupt in the same manner that objects in the lower world below the sphere of the Moon do. This is contrary to the belief of Greek and Muslim philosophers that celestial bodies are formed from the fifth element, which is distinguished by the property of being eternal and is never corrupted or disintegrates. He says, "what is said that the orbs are made of a fifth nature which is not heat, coldness, moist or dryness, is also false and has no supporting evidence."[192]

190 Ibid. 59.
191 Ibid. 66.
192 Ibid. 64.

In this respect, we remember that al-Ghazālī in his book, *Tahāfut al-Falāsifah,* refuted the claim that the Sun will never be corrupted, arguing that the Sun, "may well be undergoing a process of withering and may have been, until now, diminished by the amount of [several] mountains or more. The senses, however, would have been unable to apprehend this because estimating [such an amount] is known in the science of optics only by approximation."[193] Again, al-Bāqillānī denies that celestial objects can have any action on objects which are on the surface of the Earth, because they do not have the will necessary for taking such an action, and such an effect, if it were to occur, should be of the same kind since, "the naturally moulded to do certain actions should do so with the same effect, unlike the willing actor who can do something and its reverse."[194] Furthermore, he says, "since evidence shows that Jupiter, Saturn, the Moon and the Sun are similar bodies and, therefore, their effect should be the same."[195]

The view of the Mutakallimūn, with respect to the composition and destiny of the heavens, is completely different from the views of the philosophers in that they recognise the unity of the material world. Be it the upper world or the lower world, both are equally the same and their effects should also be the same, in that they have the same form and compositions. This requires an advanced level of comprehension indeed! This kind of comprehension formed the basis for their view of causality and causal relationships; a topic which will be presented with much more detail in Part Three of this book. At this point, I would say that it is regrettable to find some contemporary studies seemingly ignorant of the advanced views of al-Bāqillānī and al-Ghazālī.[196]

VIEWS OF IBN ḤAZM

Ibn Ḥazm was a theologian who agreed with some views of the Mutakallimūn whilst disagreeing with others. He objected to the negation of the nature of things and argues that words such as *nature* and the like were known at the time of the Prophet and used by the companions of the Prophet, with none of them denying them.[197] In this argument we observe two things:

1. that Ibn Ḥazm objected to the negation of the existence of nature as a term used to describe the properties of objects,
2. he was assuming that the *nature* of things are their observed properties.

193 Al-Ghazālī, *Tahāfut*, 49.

194 Al-Bāqillānī, *Tamhīd*, 71.

195 Ibid. 72.

196 In this respect see: André Smirnov, "*Causality and Islamic Thought*", published in "A Companion to World Philosophies", ed. E. Deutch and R. Bontekoe, Blackwell publishers, 1997, pp.493-503.

197 Fisal, 5/12

Ibn Ḥazm objected to the negation of the nature of things because he believed that once these properties are negated, things inevitably change. He says, "like the properties of wine once dropped, it would become vinegar and the properties of bread and meat once removed, they would become rubbish."[198] It is regrettable that Ibn Ḥazm did not go any further into his arguments as such a problem would have benefitted from more analysis in this context.

However, we find that Ibn Ḥazm returns to the question in another context when he discusses Divine action and expresses agreement with the views of the Ashʿaris on re-creation. We find him saying, "It is found true that at the very moment Allah changes the state of his creation into a new one, and Allah re-creates the whole world every moment without annihilating it."[199] This implies that Ibn Ḥazm agrees with the suggestion of re-creation. He supports his belief with the phrase, '*creation after creation*', which was taken from the Qur'anic verse in Surah Az-Zumar:

> *He created you from a single being, then made its mates of the same (kind). And He sent down for you eight of the cattle in pairs. He creates you in the wombs of your mothers—creation after creation—in triple darkness. (39:06)*

THE VIEWS OF AL-GHAZĀLĪ

One may expect the views of al-Ghazālī, regarding *nature,* to be along the same lines as that of al-Bāqillānī, but it turns out that this is not true. In fact, al-Ghazālī endorsed the existence of human nature in his books, *al-Iqtiṣād fi al-Iʿtiqād,* and *Ihya' Ulūm al-Dīn*. As for the intrinsic nature of inanimate things, al-Ghazālī also endorsed it, saying, "lightness is a natural force with which a body moves away from the centre [of the Earth] by nature, and heaviness is a natural force by which a body moves toward the centre by nature."[200] Also, al-Ghazālī defines moisture as, "an activated quality by which the body accepts squeezing and forming easily, and it does not pressure that but returns to its original shape and position by motion of its whole, according to its nature."[201] In fact, this definition is given nowadays for elasticity and not fluidity, but since elastic properties were understood to be caused by some sort of liquidity, they took on the same definition in this context.

Here, we see al-Ghazālī is speaking of nature and natural in terms of properties of the body rather than an active quality that can effect a change or carry out an act. One should also note that these quotations were taken from

198 Ibid. 5/16.
199 Ibid. 52.
200 Al-Ghazālī, *Iḥyā' Ulūm al-Dīn*, Beirut: Dar al-Marifa 74.
201 Ibid.

his book, *Mu'yār al-'ilm*, in which he was expressing the views of the philosophers.[202] However, to make his position on this concept clearer, I add the following quote from his later works, the famous book, *al-Munqidh min al-Dhalāl,* which exposes, without any doubts, his understanding, "The basic point regarding all of them is for you to know that nature is totally subject to God, Most High: it does not act of itself but is used as an instrument by its Creator. The Sun, Moon, stars, and elements are subject to God's command: none of them affects any act by and of itself."[203] This demonstrates that by holding that the intrinsic nature of things exists, al-Ghazālī does not necessarily mean to say that nature can act intrinsically in a way to effect changes.

The above presentation of the arguments of two prominent Muslim theologians, Ibn Ḥazm and al-Ghazālī, besides that of the famous Mutakallimūn, al-Bāqillānī, exposes the fact that they did not agree with the philosophical view of acknowledging effectiveness of inanimate matter in the happening of events in the world.

A CRITICISM FOR NEGATING NATURE

Perhaps Ibn Ḥazm and al-Ghazālī were more realistic in their acknowledgment of the intrinsic nature of objects by connecting it with their properties. In this respect, it might be worthwhile to remember that both scholars accepted the notion of non-endurance of the accidents (which is the basis of the re-creation principle) in one way or another.

The fact that we see regularities in the world demonstrates, on a practical level, the functionality of this according to given rules and laws that have been devised to exhibit natural phenomena in a consistent way. The basic, and most straightforward, thing to observe is the role of a given property in any given event. This relation between the given property and the occurrence of the event provides us with the relationship between cause and effect. Such relationships are called causal. There is no way to deny the existence of these relationships since they are ontologically acknowledged as an empirical fact and not an interpretation of phenomena. Causal relationships serve to express the permitted relationships between the variables of event accuracy in front of us. However, such acknowledgment of causal relationships does not necessarily mean that causality has to be deterministic. Such determinism needs to be proven first. This is why some geniuses of thought, such as al-Ghazālī and

202 Al-Ghazālī, Abū Ḥāmid (1911). *Mi'yār al-'ilm fī fan al-manṭiq*. Cairo: Kirdistan al-'Ilmiyyat.

203 Al-Ghazālī, *Deliverance of Error (al-Munqidh min al Dhalāl)*, translated by Richard, J. McCarthy, Boston, 1908, 10.

David Hume, had their own objections against submitting to deterministic causality and have required proof.

It is worth noting that the argument of the Ash'aris on the proportionality of effects and causes, by which they claim that a natural effective cause would induce a perpetuating effect, is no longer acceptable. Irrigating plants would not necessarily induce continuous and unlimited growth. Drinking wine induces effects different from those induced by drinking water. Eating wood, or even mud, may cause the feeling of being full, but it will not cause feeding, because wood or mud do not have same the nutritional properties as food.

The acknowledgment of the existence of certain properties or qualities in objects which makes them eligible for certain uses or effects should by no means be understood as acknowledging the independent effectiveness of those properties. If we want to amend the view of the Ash'aris' *kalām*, in this respect, we should resort to *kalām* atomism and address the subject accordingly. There should be no problem in acknowledging the intrinsic properties of objects, as long as these are accidents which endure and are under continued re-creation. Through this, we can guarantee all the requirements of the Islamic creed in respect of natural indeterminism, Divine role, and the proper understanding of causality.

Contingency, indeterminism and necessity are always at play in any natural causal event by which a new construction can be formed. The first, i.e. contingency, stands for the properties of the object which is always under continued re-creation, which cannot be determined with absolute accuracy, whilst the necessity aspect is understood as the law governing those changes. Contingency is provided by all the available properties of objects, whilst necessity is only an expression of the law by which such properties interact to produce the anticipated effects. This subject will be addressed with further details in Part Three, when I discuss the topic of causality.

SUMMARY

The principle of contingency is intimately related to the principle of re-creation. Both are based on *kalām* atomism, which assumes the special formation of the atom out of a jawāhir (substance) and *a'rād* (accidents), and all of which are based on the assumption of the temporality of the world and all its content. Each of these principles (or postulates, as they may be considered) has certain implications and consequences, and, like any theoretical framework, through their principles the upper construct of *Daqīq* al-*kalām*, can provide explanations and predictions for many problems in science and theology.

The principle of contingency has a direct implication upon the question of natural causality and the problem of Divine action. Both are related to the indeterminism of the world which makes every part and every action within it a contingent and not necessary. Necessity is manifested by the need for a law, which is normally expressed by the regularity of natural events.

The theological basis for the principle of contingency rests on the belief that God is the Sustainer of the world and, in order to establish this, He has to be the overwhelming sustaining agency in the whole world. Within this view, it might be said that the principles of *Daqīq* al-*kalām* are apologetic constructs stemming from religious belief. These judgements might be acceptable if we ignore the fact that *kalām* was not only produced to serve in defending the Islamic creed, but it was meant to develop an Islamic philosophical system of thought. That is why there were different trends and different schools of thought that came out of such developments. To say that *kalām* is meant to be a discourse in defending Islamic belief is partially true and should not be seen to be the goal.

In fact, the discoveries in modern physics at the beginning of the 20th century provided a new view of the world through the discovery of the atomic world with novel properties and logic, with wider implications. On the other hand, the Theory of Relativity has formulated a picture of the world in terms of an integrated spacetime, by which space and time form a continuum. This theory has changed our view of the world and replaced Newton's Theory of Gravity with a completely different set of concepts and explanations. Gravity, within the Theory of General Relativity has become a curvature of space time. As a consequence of this, new phenomena were discovered through the predictions imparted by this powerful new theory.

CONTINGENCY AND INDETERMINISM IN QUANTUM PHYSICS

In classical physics, every cause will necessarily have an effect. The whole world cannot be but a necessity of some kind. This is why, in classical philosophy, the world, once it existed, is deemed to be necessary and not a contingent. This is, perhaps, the basis for what Albert Einstein once queried, "whether God had any choice in the creation of the world."

In quantum physics, the matter is different. The world and all its content, and the events occurring within it are contingent and not necessary. The physical values that an object may acquire in the quantum world are normally chosen out of a set of possible values; sometimes this set is infinite. However, there is always the necessity that is imposed upon an object by the laws of physics. Such necessity is brought about by the causal relationships as defined by the laws of physics. Nevertheless, the outcome of such relationships, although highly predictable are not deterministic.

In negating classical determinism, quantum mechanics has provided us with a completely new picture of the world. Besides affirming that the world exhibits a fuzzy appearance at the sub-nuclear level, on the philosophical level, the implications of indeterminism were taken to include the objectivity and reality of our world. Events might happen, which, though unexpected, are predictable with known probabilities according to the mathematical formulations of quantum mechanics.

Being contingent and having different possibilities for the same necessity allowed for the implication that some agent is needed to make events choose one certain contingent state over the other. Such an agent is not yet known in physics, and it remains an open question.

Kalām can answer this question by suggesting that Divine actions sustain all the events of the world. In this scope, one may devise a theory of Divine action through establishing a collaboration between *kalām* and quantum physics. This is a subject which should be considered in conjunction with the new *Jalīl al-kalām*, which needs to be built up through rigorous studies. But here, I would like to point to the need for establishing the philosophical view of *Daqīq* al-*kalām* as a successful theoretical construct. As such, *Daqīq* al-*kalām* provides us with the philosophical framework needed to understand the indeterminism of quantum physics and may help provide us with better definitions for the limits that our imagination is allowed to take on objectively. On the other side of the coin, we need a better understanding of Divine action. This topic has been covered in depth during the last few decades,[204] taking into consideration, specifically, the quantum picture and the mechanism it could offer. Here, we can see how science can elevate our understanding of religion and how religion can elaborate our scientific concepts in the broad scope of knowledge and destiny.

THE ROLE OF CONTINGENCY IN THE NEW KALĀM

The principle of contingency, or indeterminism, plays a vital role in constructing the view of the new *kalām*, and providing new explanations for observed phenomena. Here, we have two subjects for which the principle of contingency can render service: the first is on the theological level, where the principle can help provide a theory of Divine action as well as explain the relationship between the world and the Divine. Here, our understanding of nature is set in service of belief, in order to enhance it and rectify it. On the other hand, the principle can be utilised to explain quantum phenomena and, perhaps, may help open the door for a new interpretation that can accommodate objectivity

204 See for example: Lameter, C., *Divine Action in Framework of Scientific Thinking: from quantum theory to divine action, Christianity in the 21st century*, 2006.

and reality in a more sensible framework. This would certainly be of great value, not only on the philosophical level, but on the scientific level too, as new concepts and understanding are necessary. Concerning the grey area in between the world and the Divine, the concept of causality stands as an important issue that needs to be addressed properly.

In traditional *kalām* there was much confusion about the subject since most Muslim thinkers thought that admitting causality in the world might imply a limited duty for the Creator. The Ashʿaris, especially, were conscious of this topic and resorted to occasionalism, which denied causality and left, in its place, the unpredictable wish of God to act as He wills. In this vision, however, the concept of the law of nature was lost, and it was hard to identify a comprehensive and consistent physical theory which could be a sufficient rival to the theory proposed by the philosophers. The occasionalist framework is loose and does not identify firm rules for confirming activities of the physical (natural) world to abide by.

The concept of custom (*ʿādah*: habit), which was offered up by Ashʿari *kalām*, was not rich enough to provide us with any sort of reliability on the part of effecting actions in the natural world. This becomes a very serious issue in a world in which one would like to know to what extent the Divine wish will take either this contingent choice, or that. The presence of a law is an important requirement for the construction of mechanisms that enable us to comprehend our world. From a philosophical point of view, and in consideration of the advancement of physical sciences, we should acknowledge the need for a deeper understanding of the laws of nature, as well as the laws of physics. Occasionalism can only be addressed in the context of the probabilistic character of physical systems; such a character which is calculable by the deterministic laws of quantum mechanics. This means that, while we admit the indeterminism of the physical world, we should also pay attention to the fact that such indeterminism keeps the choice of the contingent result unknown. That is, it defines the limits and expectations with acceptable accuracy. This is certainly a valuable part of the determinism of the laws of physics.

Chapter Seven

The Integrity of Space and Time

In their views on space and time, Muslim philosophers, such as al-Fārābī, Ibn Sīnā, and Ibn Rushd, mostly followed the Greek philosophers; namely Plato, Aristotle and Plotinus. The Mutakallimūn followed another approach; they constructed their views mainly from the Qur'an, the prime source of Islam. The *kalām* views were different from those of the philosophers in some fundamental respects. The Mutakallimūn presented their views about space and time when discussing a number of fundamental issues in religion and natural philosophy, most important of which were the problems of creation and motion. Most of the Mutakallimūn considered time to be discrete, composed of non-divisible units called *ān*, meaning an "instant." In conforming with their atomic theory, the Mutakallimūn viewed the motion of a particle as composed of a finite (discrete) number of transitions over a trajectory separated by stationary points. This concept was fundamentally different from the conventional Aristotelian concept of translational motion, which was described as the transition from one place to another during a given duration of time.[205]

205 See for example: Peter K. Machamer, *Aristotle on Natural Place and Natural Motion*, Isis, Vol. 69, No. 3 (Sep, 1978), 377-387.

The point to note here is that the Mutakallimūn described space and time as being an integrated entity that is manifested in the occurrence of the event. Moreover, they considered both space and time to be on relative scales, as they are always addressed in comparison to other references, thereby, rejecting the notion of absolute space and absolute time.

In addition to describing the views of the Mutakallimūn about space and time and, inevitably, motion, I will consider in this chapter, the views of two of the great traditional scholars of Islam who did not formally subscribe to the schools of *kalām* but who, nonetheless, frequently espoused some of the doctrines of *kalām* in their arguments, despite having different views about certain other matters. They are Ibn Ḥazm al-Ẓāhirī and Abū Ḥāmid al-Ghazālī. I have chosen these two thinkers because they represent, perhaps, the highest level of traditional Muslim intelligentsia, and expressed *kalām* theories in a theological context as well as theological concepts within the framework of *kalām*.

Ibn Ḥazm, who was born and lived in Córdoba (Spain), expressed most of his philosophical views in his famous book, *Kitāb al-fiṣal fī al-milal wa al-ahwā' wa al-niḥal*, in which he discussed philosophical thoughts and views of many religious groups and factions. Primarily, he stressed the importance of sense perception, asserting that human reason can be flawed. This contradicts the doctrine of al-Ghazālī. While recognising the importance of reason and acknowledging that the Qur'an encourages rational reflection, al-Ghazālī believed that this reflection was concerned mainly with revelation and the senses. Thus, it is a form of sensory reminder to admire the glory of God. Accordingly, he concluded that reason is not to be taken necessarily as a faculty for independent research or discovery, but that sense perception should be used in its place, an idea that sounds like a forerunner for empiricism. Although, Ibn Ḥazm did not subscribe to any of the *kalām* schools, despite his critique of the Muʿtazilis and Ashʿaris, it is not difficult to see that he used some of their thoughts in his arguments. This, obviously, stems from having a common base with those arguments, which were, of course, based on Islam.

Al-Ghazālī, the most famous Muslim intellectual and thinker, lived and taught in Baghdad at the Niẓāmiyyah School during the last two decades of the 11th century. Al-Ghazālī, too, did not officially subscribe to either of the two main schools of *kalām*, but he certainly used their arguments in his book, *Tahāfut al-falāsifah* (*The Incoherence of the Philosophers*). In his arguments, he used the concepts of *kalām* extensively, and added much elaboration and ingenuity to those concepts. Al-Ghazālī shared some of the views of Ibn Ḥazm, and sometimes used the same arguments for whichever problem was under discussion.

In this study, I will present the views of both scholars, revealing their most important ideas in an attempt to demonstrate an important part of the Islamic view of space and time. I do not intend to set the discussion in a historical context, nor will I present a history of thought on the concepts of space and time, but, rather, I will concentrate on presenting the ideas and views of these two Muslim thinkers in the context of *kalām*. Whenever necessary, I will also discuss the opinions of other theologians and philosophers in order to briefly cover their basic thinking. In order to provide a full coverage of the concepts and thoughts to date, I will also present the views of the two main theories of the 20th century; namely Einstein's relativity and quantum mechanics. These are presented as references for comparison and to assess the richness of traditional Muslim scholarship, which I will leave the reader to appreciate.

SPACE AND TIME ACCORDING TO ISLAMIC KALĀM

Perhaps the best definition of space, according to Islamic *kalām*, is the one given by al-Jurjānī, "the conceived empty place which is occupied by the body and in which its dimensions are extended."[206] This implies that space is an imagined place that does not exist unless it is occupied by a body. To clarify this position, we need to explain another term directly related to the concept of space, which is "occupancy" (*taḥayyuz*). This term has been considered before in Chapter Three, but we will deal with it here again. According to al-Jūwaynī, *ḥayyīz* is "the place for an envisaged *jawhar*."[207] In this respect, we should remember that, according to most of the Mutakallimūn, *jawhar* (see Chapter Three) has no size or area. However, as for the other concept of empty space (*khalā'*), the Mutakallimūn considered it to mean "the space which is left behind when the occupying body is removed."[208] So, the empty space should exist. Without it, no motion could be achieved. Ibn Matawayh, a famous Muʿtazili, presented an argument in favour of the existence of *khalā.'*[209] These concepts were commonly understood by Ashʿaris and Muʿtazilis, although they had different views on some of the finer details.

SPACE AND TIME IN MAQĀLĀT

Little can be construed from the content of the *Maqālāt* of Al-Ashʿari about the concepts of space and time. Al-Ashʿari refers to different expressions for the meaning of *space* without mentioning known sources. He says: [210]

---

206 Al-Jurjānī, *al-Tarīfāt: al-Makān.*

207 Al-Jūwaynī, *al-Shāmil*, 156.

208 *Taʿrifat, al-khalā'*

209 Tadhkirah, 116–124.

210 Al-Ashʿari, *Maqālāt al-Islāmiyyīn*, 329.

1. some said the place of a thing is its substrate, in which it stays.
2. some others said the place is what surrounds a thing.
3. and some said the place of a thing is what prevents it from falling
4. some others said the place of things is the space (the air or atmosphere).
5. some others said the place of a thing is its destiny.

As for the meaning of *time* al-Ashʿari reports the following:[211]

1. Abu al-Hudhayl said that time is the difference between events, and it is the duration between one event and the next.
2. Al-Jubba'ī said that time is what you designate for a thing, such as saying, 'I will come once Zayd arrives,' so here you have made the coming of Zayd a time for your arrival. Al-Jubba'ī further claims that time is [recognised by] the motion of celestial objects as God has designated.
3. Some others said that time is a Transient *(a'rād)* and we cannot say what it is or know its reality.

The concepts of space and time in Islamic *kalām* are very much connected with the principle of discreteness (atomism) and the principle of re-creation, which was developed by the Mutakallimūn to apply to all discrete entities and, specifically, the concept of space and the question of occupancy (*taḥayyuz*) of this space. Space, as well as time, was conceived as being discrete. Al-Jurjānī defines time as, "a known renewable by which an envisaged unknown is estimated."[212] Clearly, in this definition we can spot two features given to time: the first is intrinsic; being renewable, and the second is functional; connecting, simultaneously, two events. Perhaps one might say that the second feature allows this definition of time to include simultaneity, since the conjunction of two events requires the defining of two times, thus involving simultaneity. However, this was not so clear in the concept as defined by the Mutakallimūn.

The discreteness of space is another subtle concept, which might not be as clear as in the case of time. However, this discreteness might be clarified by pointing to how the Mutakallimūn added two or more jawāhir together. In this case, they denied that two jawāhir might diffuse into one another, but insisted that they could only touch each other (*tamās*). Thus, when two jawāhir are attached to each other, a line is formed; to make a two-dimensional surface, we would need four jawāhir, and to make a three-dimensional volume, we need eight. According to *kalām*, this describes the fundamental construction of an extension in space. This exposes an abstract understanding of the basic elements that constitute matter.

211 Ibid.

212 Al-Jurjānī, *al-Taʿrīfat: al-Zaman*.

In this respect, Max Jammer raised the question of whether it was sheer coincidence that Leibniz suggested *monadology* to sketch the metaphysics of simple substances, or "*monads*." Jammer presents an argument pointing to the possibility that Leibniz may have adopted the atomic theory of *kalām*. On the other hand, he says that, "consequential thought led the *kalām* to the conclusion that space, as well as matter (and time), is of atomistic structure."[213]

Related to the concepts of space and time, we naturally come to the concept of motion. Since space and time were taken to be discrete, according to *kalām*, motion becomes a discontinuous process. Motion is viewed as a series, or sequence, of momentary leaps. As the jawāhir occupy different individual places in succession, physical motion has to be discontinuous. Jammer presented a beautiful argument affirming the discreteness of space according to *kalām*, which goes:

> The discrete structure of space can be inferred from two premises (1) the discreteness of time and (2) the Aristotelian inference from the continuity of space to that of motion, and from the continuity of motion to that of the continuity of time. Since the latter, according to the first premise is denied, the formal application of the *modus tollens* leads to the conclusion that space is not continuous.[214]

Jammer claims that Galileo's discussion of the problem of discrete motion in his *Discorsi e Dimostrazioni Matematiche Intorno a Due Nuove Scienze* (Discourses and Mathematical Demonstrations Relating to Two New Sciences), and his treatment of the infinite and the indivisible, is reminiscent of the ancient teachings of *kalām*.[215] Apart from this, very little is known about the influence of the *kalām* conception of space and time on scholastic thought in medieval Europe. But, since it is well established that the works of al-Ghazālī and Maimonides, with their references to the atomistic space theories of *kalām*, were widely read by scholars, Jammer seriously questions the possibility that this atomistic theory of space could have escaped their attention.[216] This is a very important question indeed, taking into consideration the influence of discreteness on both Leibniz and Galileo. However, in order to resolve this question, we would need to invest a great deal of impartial effort into analysing this scholastic legacy. Perhaps this is part of the homework that modern Muslim scholars specialising in the history of science should do.

213 Jammer, Concepts of Space: The History of the Concepts of Space in Physics. (2nd ed.), Cambridge, MA: Harvard University Press, 65.

214 Ibid

215 Ibid., 67

216 Ibid., 68

Space and time are two entities that are essential for our understanding of the physical world. Space seems to be more real than time, as it is objectively present; it is part of what we see around us. Time is not so tangible to our senses, as it seems to be less objectively present. For us to feel the presence of time, events need to occur, and our world cannot endure two instances without something changing. It was thought that time exists like a river running independently of any concern for those at its banks; it can only be affected by the topography of the land through which it passes, running quickly as the land slopes and slowly as it climbs a hill. Similarly, time is affected by the topography of space through the mutual interaction that keeps the path of light intact.

Time seems to have only one direction, regardless of the many mathematical formulae that allow for time reversal. Nature, through the requirements imposed by thermodynamics, prevents time reversal and that is why we have a time arrow. Translation in space can take forward or backward directions, i.e. we can retrace our path as we move, but can never retrace our past time. It would appear that time reversal is a fiction.

SPACE AND TIME ACCORDING TO IBN ḤAZM

Ibn Ḥazm was born into a rich and influential Córdoban family. He received a distinguished education in religious sciences, literature, and poetry. Profoundly disappointed by his political experiences and offended by the conduct of his contemporaries, Ibn Ḥazm subsequently left public life and devoted his last thirty years to literary activities. In this time, he produced a reported 400 works, of which only forty still survive. He covered a range of topics, including Islamic jurisprudence, history, ethics, comparative religion, and theology, as well as producing his famous work, *The Ring of the Dove*, on the art of love. Ibn Ḥazm was a leading proponent and codifier of the Ẓāhirī School of Islamic jurisprudence. *The Encyclopaedia of Islam* refers to him as having been one of the leading thinkers of the Muslim world and he is widely acknowledged as the father of comparative religious studies.[217] In his treatise, *Kitāb al-Fiṣal fi al-milal wa al-ahwā' wa al-niḥal*, Ibn Ḥazm stressed the importance of sense perception. While he recognised the importance of reason—since the Qur'an itself invites reflection—he argued that this reflection mainly refers to revelation and sense data, because the principles of reason are themselves derived entirely from our sense experience. He concluded that reason is not a faculty for independent inquiry, research, or discovery, but that sense perception should be used in its place, an idea that forms the basis of empiricism.[218] In this

217 *Encyclopaedia of Islam*, 2nd ed., s.v. "Ibn Ḥazm"

218 Islamic philosophy online, Ibn Ḥazm. Translated by Miriam Rosen. Accessed 27 June 2015. <http://www.muslimphilosophy.com/ Ḥazm/ibnḤazm.htm>.

argument, Ibn Ḥazm was, perhaps, criticising the Greeks, who were known to have stressed the value of reason and considered works, without much need for experiment.

In what follows, I will present Ibn Ḥazm's views on space and time as he has presented them; mainly in *al-Fiṣal.* First of all, we should know that Ibn Ḥazm refused to acknowledge the existence of any physical infinite, including space and time. He tried to refute the infinite extension of time by simple logical arguments. He said, "Everything that exists in reality is confined by number, countable by its own nature, and by [the term] nature we mean the force in the thing by which its properties are run." Then he said, "Everything that is confined by number would be countable by its own nature, therefore it is finite, and consequently the world is finite."[219] Accordingly, he refused to accept the existence of anything which had an infinite extension:

> An infinite would in no way exist in reality, and whatever exists only after an infinite regress could not exist at all, because being "*after*" necessitates finiteness, and an infinite has no "*after*." Consequently, nothing can exist after something in infinite regress and, since things do exist one after the other, therefore all things are finite.

So, this is how Ibn Ḥazm thought of the impossibility of infinity. Thus, Ibn Ḥazm makes me think that, in fact, he was adopting the doctrine of the finite divisibility of things, despite his denial of *kalām* atomism, which was based on his theological argument that Allah is able to divide things infinitely. Otherwise, as I have shown in Chapter Three Ibn Ḥazm agreed that a non-divisible entity may exist in reality but not in theory.

Ibn Ḥazm defined time as, "the duration through which an object stays at rest or in motion, and if the object is to be deprived of this [rest or motion] then that object will cease to exist, and time will cease to exist too. Since the object and time both do exist, therefore they both co-exist."[220] Clearly, in this definition, Ibn Ḥazm associated the existence of time with the existence of the body, which pointed to the connection between space and time. Following on from this, he argued that the time of the world has finite duration as well as a beginning, "Any object in the world and every accident associated with it and time is finite and has a beginning. We see this sensibly and objectively because the finiteness of an object is obvious through its size and through the time of its existence."[221] Therefore, objects of the world are finite and, in the above sentences, it is clear that Ibn Ḥazm denied the existence of anything infinite. But,

219 Ibn Ḥazm, *Fisal*, 57–58.
220 Ibid., 61.
221 Ibid., 57.

in the case of time, he went even further to consider time as being composed of finite instances, moments that pass one after the other:

> The finiteness of time happens, although what comes after that which has passed, and the exhaustion of every time [period] after its existence, as "now" is the limit of it, and it is this [now] which separates the two times: the immediate past and the immediate future, and it is as such that one time ends and another would start. And every period of time is composed of finite times that have beginnings.

This makes it clear that Ibn Ḥazm considered time to be discrete on the ontological level, despite his general denial of the "indivisible parts." Ibn Ḥazm used these concepts to argue that God existed in neither space nor time:

> Because God is not occupied with time and has no duration or end, because time is the motion of whatever is timed, its motion from one place to another, or its duration when at rest in one place, and God is neither in motion nor is at rest and [there is] no doubt that He is not timed and has no duration or end, and He is originally not in one place.[222]

Additionally, he says:

> God is not [confined] to a time and has no duration, because time is the motion of any timely object and its transition from one place to another or the duration of its stay at rest in one place, and God neither is movable nor is at rest.[223]

It should be noted from the above quotations that Ibn Ḥazm understood time to run sequentially. He thought that the passage of time occurs in sequential moments, one after the other; as the moment passes, it becomes the past. Subsequently, a new moment replaces the old one, and so time passes on.

Although Ibn Ḥazm did not seem to accept the principle of discreteness as envisaged by the Mutakallimūn, his writings express a belief that time is divided into finite instances, a notion very similar to time discreteness. Indeed, for Ibn Ḥazm to have been consistent in his views about space, time, and the creation of the world, he should certainly have adopted the discreteness and finiteness of parts. Those who claim the contrary should read all the available descriptions and analyse all of his arguments, not only in their expressions but in their consistency, to see that he could not but have adopted discreteness, despite his refusal to accept the notion of a non-divisible part on a theological basis.

222 Ibid. 62.

223 Ibid., 22

Concerning space, it seems that Ibn Ḥazm adopted the Aristotelian view to define space and argued for the absence of voids. In *al-Fiṣal,* he defined space in a similar way to Aristotle, "Because the space that we know is the place surrounding the body localised within."[224] It is also notable that Ibn Ḥazm refutes the existence of absolute space. He refers to a discussion he had with a group of people:

> They say that absolute space and absolute time is not what we have defined previously, because space and time are changing, and it would suffice to refute their argument of defining an unaccustomed concept of space and an unaccustomed concept of time without having evidence for it.[225]

This absolute space, he described, as being the void that exists independent of objects or bodies. In order to refute this claim, Ibn Ḥazm used lengthy dialectical arguments, which I find are not very convincing. A similar discussion was considered by al-Ghazālī, but with more sophisticated arguments.

SPACE AND TIME ACCORDING TO AL-GHAZĀLĪ

Al-Ghazālī was a prominent thinker who produced such a colourful range of thoughts that it is a puzzling task to associate him definitively with any one school of thought, other than to say that he belonged to a special school of his own. He might be considered an Ashʿari theologian, a philosopher, or a Sufi monk. He expressed a multitude of thoughts in his writings and was experienced in all the possible methodologies of his time.

Al-Ghazālī viewed space and time as being two entities that should be treated on an equal footing. His best presentation on this subject can be found in his treatise, *The Incoherence of the Philosophers,* where he refutes the philosophers' claim about the eternity of the world. Al-Ghazālī presented similar arguments about time to those of St. Augustine, some of which I have mentioned in Chapter Eleven (Astronomy and Cosmology). However, one can confidently say that, in presenting these arguments, al-Ghazālī was also speaking as a representative of the Mutakallimūn, since he was using their dialectical method, and concepts about space and time. It would also be fair to say that al-Ghazālī presented these arguments with much originality and thought. He used an analogy of space and time, the *above* and the *below* versus the *before* and *after,* in order to proclaim an equivalence between spatial and temporal extensions.

Al-Ghazālī considered time to have been created alongside the world, and not before it, "Time is originated and created, and before it there was no time

224 Ibid. 73.
225 Ibid. 61.

at all. We mean by our statement that God is prior to the world and time: that He was, and there was no world and that then He was, and with Him was the world."[226] In response to the question about the time that passed before the creation of the world, al-Ghazālī presents an analogy of space, whereby he says that when we talk about the world as a whole, we should realise that there is nothing beyond its surface:

> Similarly, it will be said that just as spatial extension (*buʿud makānī*) is a concomitant of a body, temporal extension (*buʿud zamanī*) is a concomitant of motion. And just as the proof for the finitude of the dimensions of the body prohibits affirming a spatial dimension beyond it, the proof for the finitude of motion at both ends prohibits affirming a temporal extension before it, even though the estimation clings to its imagining it and its supposing it, not desisting from [this]. There is no difference between a temporal extension that in relation [to us] divides verbally into 'before' and 'after,' and a spatial extension that in relation [to us] divides into 'above' and 'below.' If it is legitimate to affirm an 'above' that has no above, then it is legitimate to affirm a 'before' that has no real before, except an estimative imaginary [one], as with the 'above.'[227]

The most important piece of information in the above quotation is the reference made by al-Ghazālī to the term "temporal extension" alongside the term "spatial extension." This was something new for the intellectual era in which al-Ghazālī lived and, indeed, it reflects a deep understanding of the meaning of space and time in our real world, and the reason for their absence before the creation of the world.

Al-Ghazālī continued arguing about the relativity of the 'before' and the 'after', responding to criticism that may have been directed against his analogy of time with space, where it could be said that space and time cannot be treated on an equal footing:

> This comparison is contorted because the world has neither an "above" nor a "below", being, rather, spherical, and the sphere has neither an "above" nor a "below." Rather, if a direction is called "above" this is inasmuch as it is beyond your head; the other [direction is called] "below" insofar as it extends beyond your foot.[228]

Al-Ghazālī counteracted this by saying:

> This makes no difference. There is no [particular] objective in assigning the utterance "above" and "below," but we will shift to the expressions

226 Al-Ghazālī, *Tahāfut*, 31.
227 Ibid. 31.
228 Ibid. 33-34.

> "beyond" and "outside" and say, "The world has an inside and an outside: is there, then, outside the world something which is either filled or empty space?" [The philosophers] will then say, "Beyond the world there is neither a void nor filled space. If by 'outside' you mean its outermost surface, then it would have an outside; but if you mean something else, then it has no outside." Similarly, if we are asked, "Does the world have a 'before'?" we answer, "If by this is meant, 'Does the world's existence have a beginning, that is, a limit in which it began?' then the world has a 'before' in this sense, just as the world has an outside on the interpretation that this is its exposed limit and surface end. If you mean by it anything else, then the world has no 'before,' just as when one means by 'outside the world' [something] other than its surface, then one would say, 'there is no exterior to the world.' Should you say that a beginning of an existence that has no 'before'" is incomprehensible, it would then be said, "A finite bodily existence that has no outside is incomprehensible: If you say that its 'outside' is its surface with which it terminates, [and] nothing more, we will say that its 'before' is the beginning of its existence which is its limit, [and] nothing more."[229]

Al-Ghazālī's continued defence of his concepts of space and time led him to question the size of the universe and whether it could have been originally created larger or smaller than its known size. This is the challenge that I will address in the next chapter. Therefore, I conclude that al-Ghazālī, by adopting the basic views of *kalām*, considered space and time to be on an equal footing and was able to envisage the relationship between them in a way that perceived their relativity in the sense of how it is accounted for by the observer conventionally (obviously though, not in Einstein's sense, which accounts for the relativity of time more profoundly). Accordingly, he was able to present the reason for there not being an arrow of time before the creation of the world. These ideas of al-Ghazālī are important because they express the basic concept of time, as we now know it, after the elaborations given by the Theory of Relativity.

SUMMARY

In summary, I can say that the most important finding of this study is that the Mutakallimūn understanding of space and time differed in some important aspects from the understanding adopted by most of the prominent Greek philosophers. The main aspects of the *kalām* understanding of time and space include:

1. Considering space and time as being created alongside matter upon the creation of the world. Thus, before the creation of world there was no space and no time.

229 Ibid. 35.

2. Time being discrete and composed of indivisible parts.
3. Addressing both space and time as being relative measures, depending on the reference point, rather than being absolute.
4. Treating time functionally—they mostly dealt with astronomical time, which they considered a measure of motion, or change, in general.
5. Negating the existence of absolute space and absolute time.

The importance of this understanding of time as being a discrete entity provides us with a possible unification of this concept with discrete space. This could help, philosophically, in evaluating a possibility for unifying important parts of our current physical theories. For example, the unification of gravity and quantum mechanics requires that we envisage some kind of discreteness of the spacetime interval, a long sought-after concept that would enable quantising spacetime. Another point, which might be of some value, is the Mutakallimūn's refusal to adopt the concepts of absolute space and absolute time; concepts which, for a long time, had formed part of the basic assumptions of classical physics. Historically, physics demolished the concepts of absolute space and absolute time in 1905, when Albert Einstein discovered the physical relativity of space and time.

For sure, one cannot claim that the Mutakallimūn advocated a comprehensive theory of space and time, since they could not continue their vision in constructing a consistent theory of motion, but certainly the above aspects of their understanding are worth studying and developing, at least on a philosophical level.

Part Three
Applications

In Part Two of this book, I outlined the main principles of *Daqīq* al-*kalām*. These principles, which were construed from the original doctrines of *kalām* with respect to their views regarding natural philosophy, will now be taken to furnish the basis for the modern Islamic view of physical world. They provide us with the necessary guidance to formulate our views within a consistent theoretical framework when considering the fundamental questions of modern natural philosophy. The claimed theoretical framework is composed of these principles, and their utilisation in constructing the views through studying the concerned problems will rigorously contribute to the formation of the new construct for natural philosophy. Since the principles of *Daqīq* al-*kalām* are not directly related to theological doctrines, it would be useful for non-Islamic studies to consider it and utilise its implication in their views concerning natural philosophical questions.

In this part of the book, I will discuss some fundamental questions, using examples of the principles of *Daqīq* al-*kalām* in the analysis and in constructing the views of each case. Where available, I will present the historical views of the Mutakallimūn, alongside quotes from the prime source of knowledge in Islam, the Qur'an. These questions are practical examples that show how the new *kalām* deals with issues concerning natural philosophy. The presented analysis and studies are by no means exhaustive and there is room for more in-depth investigations, which could possibly cover a wider scope and bring in further arguments in each case.

In Chapter Eight, I present the proposal of re-creation which sets a very fundamental view of *Daqīq* al-*kalām* to the physical world. This proposal is construed from the principle of re-creation originally set by the Mutakallimūn and was the subject of Chapter Five. By setting this proposal in the terminology of modern physics, I am able to analyse and explain phenomena which occur in the quantum world. In fact, the fundamental problem of measurement in quantum mechanics gain a better resolution under the proposed scheme of re-creation, while many other questions and phenomena are better explained.

In Chapter Nine, I discuss the problem of motion, in which I specifically combine the principle of atomism with the principle of re-creation in order to resolve two problems at the same time; the old problem of *ṭafra*, which was negated by most of the Mutakallimūn, and the contemporary problem of defining the trajectory of a quantum particle passing through two adjacent slits. This might help offer a resolution for the problem of understanding the bizarre dual behaviour of quantum particles.

In Chapter Ten, I present the old-age problem of causality. Here, I expose the fact that the pioneering Mutakallimūn, including al-Ghazālī, did not deny natural causality but that some of them, specifically the Ashʿaris, may have denied human causality. On the other hand, al-Ghazālī appears to deny deterministic natural causality while affirming causal relationships; defining them as incidents that occur as a custom expressing the natural behaviour. It will be shown that most of the Muʿtazilis followed an approach similar to that of the Ashʿaris. However, they have different positions on the question of human causality. In this chapter, I propose a new view of causality and, perhaps, deliver a clearer view about natural causality with an Islamic perspective, which combines the principles of re-creation and contingency to offer, what may be, a practical solution for the problem.

In Chapter Eleven, I present the views of the Mutakallimūn with respect to a number of astronomical problems. Most of the chapter is devoted to discussing the debate between al-Ghazālī and Ibn Rushd (Averroes) on astronomical and cosmological questions. In particular, I present their views on the question of whether the universe may originally have been larger or smaller than it currently is, and on the question of the post-eternity of the Sun and the possibility of its degeneration. The analysis and discussion of these two questions is a bold example that refutes the claim that al-Ghazālī was "forcefully and masterfully rebutted by Ibn Rushd."[230]

In Chapter Twelve, I discuss the anthropic principle and present a view that offers a possible explanation for the observed fine tuning of the fundamental constants and physical quantities contained within the universe. The proposed idea goes beyond recognising the fact that fine tuning is necessary for our existence, and deals with the basic laws which govern the world.

In Chapter Thirteen, I consider the problem of biological evolution, and show that the Qur'an, contrary to common belief, does not object to the notion of evolution. In fact, the Qur'an contains signs which indicate that man had passed through several stages of evolution on his way to acquiring his proper shape and adequate capabilities, before receiving the blessing of his Creator, who, by giving him the *holy breath,* endowed him with intellect and the human spirit. A discussion of some fundamental aspects of the neo-Darwinian and non-Darwinian evolutionary schemes is also included in an endeavour to present a new view that might allow for philosophical contemplation about this fundamental topic in light of the principles of re-creation and contingency.

Finally, in Chapter Fourteen I present some suggested problems suitable to be considered by researchers and post graduate students. Such problems are only examples for the new tends in *Kalām* research and certainly the interested research leader can provide many other examples for serious research work in developing the new *Kalām* in natural philosophy.

---

230 Guessoum, *Islam's Quantum Question*, 76.

Chapter Eight

The Re-Creation Proposal

This proposal was presented at a conference organised by the British Science and Religion Forum, and held at Liverpool Hope University. The presentation was published in *Matter and Meaning*,[231] with a more technical version of it being available at www.arXive.org.[232] The reason for considering this article here is to enable the reader to understand the basis of the calculations performed in the next chapter as regards to discrete motion, as well as the frequent usage of the principle of re-creation in subsequent chapters.

On the theological side, quantum indeterminism is one of the main pillars of the argument in favour of quantum Divine action. This argument exploits the fact that events in nature can only be determined within a limited certainty, and that the occurrence of these events is probabilistic in nature. Christopher Lameter[233] investigated this subject with the aim of justifying his belief in a God who acts with consideration for the scientific framework of

231 Altaie M. B. "Re-Creation: A Possible Interpretation of Quantum Indeterminism," In *Matter and Meaning*, edited by Michel Fuller, 21–37. Newcastle Upon Tyne: Cambridge Scholar Publishing. (2010).

232 Altaie, M.B. (2009) Re-Creation: A possible interpretation of quantum indeterminism, www.arXive.org, arXiv: 0907.3419 v2 [quant-ph].

233 C. Lameter, *Divine Action in the Framework of Scientific Knowledge*, (Christianity in the 21st century 2005).

quantum mechanics. Since a concept involving Divine action was especially relevant to theology, Lameter believed that a theory of Divine action, which was compatible with contemporary physics, was a fundamental requirement for the credible consideration of how God could act in the framework of a contemporary worldview.

It is a common understanding among physicists that the concept of quantum measurement is still a problem requiring a clarification of the deep implications of quantum theory. There is no consensus among physicists: instead, we have many different views regarding how quantum measurements can be interpreted. Quantum measurement is the backbone of applied quantum mechanics and, therefore, is necessary to resolve this problem for the further development of quantum theory.

In this chapter, I will present a new interpretation of quantum measurements based on the notion of continued re-creation of the physical properties of systems. Using this, I will try to explain some of the basic principles of quantum mechanics on a conceptual level only, thereby avoiding mathematical details. Then, I will try to foresee some of the physical implications of such an interpretation and will glimpse upon the philosophical and theological implications.

EARLY DEVELOPMENT OF QUANTUM THEORY

The discovery of the wave properties of particles, the particle properties of waves, and the discreteness of many observables in the atomic realm has established the need for a new description of entities of the microscopic world. At the beginning of the 20th century, many basic problems in atomic physics were addressed leading to the establishment of quantum mechanics as a paradigm to explain the observed properties of the atomic realm. The most fundamental notions of early quantum mechanics were based on the assumption that particles behave like waves. The main difficulty in designating a wave-like description for particles lies in the fact that particles are localised, whereas waves are extended. This problem was overcome by Louis de Broglie, who suggested that a particle can be represented by a plane-wave which has a wavelength inversely proportional to its momentum. This notion was utilised to describe particles in terms of de Broglie waves, with the wavelength being that of the group of waves representing the particle. This description opened the way to formulating classical localised particle-mechanics in terms of wave-mechanics. Accordingly, a wave equation was devised by Erwin Schrödinger in 1926 to describe the time development of atomic particles in a field of force.[234] The

234 E. Schrödinger, Ann. Phys. (Leipzig) 79, (1926): 361.

need to consider the spin of the electron and the Lorentz invariant equation of motion, required the introduction of a special relativistic formulation of the problem, and led to the well-known Dirac equation for the electron, which was discovered two years later.[235]

In essence, the wave-like description of atomic particles demonstrates how they display all the properties of a wave phenomenon. Particles, such as atoms and electrons, are now identified as 'quantum states' and symbolised by the wave function $\psi(x,y,z,t)$. This is a mathematical expression summarising the physical content of a physical system in terms of spacetime coordinates and other parameters of the system, such as energy and momentum. The mathematical nature of ψ(x,y,z,t) was recognised from the early days of formulating Schrödinger's equation, when it was realised that the wave function has no direct physical meaning in itself. However, Max Born[236] the German physicist identified |ψ(x) |2=|ψ* (x)ψ(x)| as standing for the probability density[237] of finding the particle in position *x*.

The wave-mechanical description of particles set by Schrödinger was best realised by saying that a particle is a wave-packet that is composed of superposing basic (plane) waves. This simple description, however, faced some significant difficulties. The slightest dispersion in the medium will pull the wave-packet apart in the direction of propagation. Even without such dispersion, it will always spread more and more in the transverse direction. Because of this blurring, a wave-packet does not seem to be a very suitable way to represent a particle.

In the early summer of 1925, shortly before Schrödinger formulated his wave equation, Werner Heisenberg[238] conceived the idea of representing physical quantities using sets of complex numbers. This was elaborated into what has become known as "matrix mechanics," the earliest consistent theory of quantum phenomena.[239] Both views, i.e. the wave-mechanics of Schrödinger and the matrix mechanics of Heisenberg, are said to be equivalent, despite there being differences in some basic concepts and formulation.

---

235 P. A. M. Dirac, Dirac, P. A. M. "The Quantum Theory of the Electron." *Proceedings of the Royal Society of London*. Series A, 117 (778): 610–624.

236 M. Born, (1926). Zur Quantenmechanik der Stoßvorgänge, *Zeitschrift für Physik* 37 863-867.

237 The star (*) symbolises complex conjugate and the total probability is the integral all over the allowed space.

238 W. Heisenberg, Über quantentheoretische Umdeutung kinematischer und mechanischer Beziehungen, *Zeitschrift für Physik*, 33, 879-893 (1925); Math. Ann. 95, (1926): 683.

239 M. Born, W. Heisenberg and P. Jordan, (1925). *Zur Quantenmechanik II, Zeitschrift für Physik*, 35, 557-615. 1925

Some years later, Jon von Neumann[240] showed that quantum mechanics could be formulated as a calculus of Hermitian operators in Hilbert space. The wave function was represented by a complex vector in an infinite-dimensional space covered by basis vectors. According to von Neumann's formalism, a physical system is completely described by the state function $| \psi >$, which is a vector in an infinite-dimensional Hilbert space. A measurement of any observable a belonging to the system is the result of the action of a mathematical operator $\hat{A}$ corresponding to that observable, on the wave function representing the system. The result of the operation is to produce a number (called the *eigenvalue*) that stands for the observable at the moment of measurement. On this new comprehension, natural objects, which were objectively identified as ontologically existing, came to be seen as new epistemological entities represented by abstract mathematical forms. It should be emphasised that this is a very important turning point in the history of scientific thought. The fact that $| \psi >$, which represents the physical system, is a mathematical expression that has no direct physical meaning (as noted earlier), and the fact that physical observables became obtainable in the theory only as a result of operating certain mathematical operators on $| \psi >$, is surely a clear indication of the fundamental turn that was implied by quantum mechanics.

The eigenvalue of an operator cannot be taken to stand for the physical value of the observable; it has to be averaged within the state of the system, after which it is called the *expectation value* of the operator $\hat{A}$ when the system is at the state $| \psi >$. This value is the average of all possible measurements that can be carried by the system in the state $| \psi >$.

THE HEISENBERG UNCERTAINTY PRINCIPLE

To the surprise of many physicists, some aspects of the wave-like description of particles led to uncertainties when determining simultaneous pairs of observables; called complementary observables, such as position and momentum, or energy and time. This was expressed by Heisenberg's uncertainty principle which, in one of its forms, states that the position of a particle and its momentum can never be determined simultaneously with absolute accuracy.[241] Mathematically, this is expressed as $\Delta p \Delta x \geq \hbar/2$ where $\hbar$ is Planck's constant. This principle contributed to the indeterminacy of the quantum world and has attracted much attention and interest from the world of physics.

240 J. von Neumann, *Mathematical Foundations of Quantum Mechanics*, trans. Robert T. Beyer (Princeton, New Jersey: Princeton University Press, 1955).

241 See for example: Heisenberg, Werner. *Physics and Philosophy: The Revolution in Modern Science*. New York: Prometheus, 1999.

In Schrödinger's picture of wave mechanics, the uncertainty principle is a consequence of the wave-mechanical description of particles, whereby, when a particle is represented by a wave, its position and momentum are, inevitably, distributed. The wave number which corresponds to the momentum of the particle becomes a distribution in momentum space and the position becomes a finite width of the wave packet. The Fourier analysis of such a description shows that the wave description requires some non-locality of position, which leads to an inherent uncertainty in these variables, i.e. in position and momentum. A similar situation is found when measuring time and energy, where it leads to a mutual uncertainty between the time interval and the associated energy. In this case we have: $\Delta E \Delta t \geq \hbar/2$

In matrix mechanics, we have three basic entities: the operators, the states and the observables. Generally, the operator is a mathematical expression which acts on the wavefunction (or state vector) representing a physical system, transforming it into another state. If the new state is just a scaling of the original one, then the function is called an *eigenfunction* and the scale factor is called the *eigenvalue.* This eigenvalue corresponds to the physical observable associated with the respective operator when acting on a specific eigenfunction. Generally, operators are square matrices that, unlike numbers, do not commute. For example, if p is the momentum and x the position operator, then $px \neq xp$. This means that if the operator x was to act on the function and was then followed by the action of p, the result would not be the same as if p acted first, followed by x. Thus, we say that the operators of position and momentum do not commute. This was the discovery of Werner Heisenberg in 1925.[242] Later, it was found that the same was true of the operator for time and the Hamiltonian (the operator corresponding to the energy in a system).

It is important to note that the discovery of the indeterminacy of position and momentum caused a tremendous shock in classical physics. The classical equation of particle motion requires knowing both its initial position and initial momentum. Having been denied such knowledge, physicists were puzzled by the solution of the equation of motion, which caused the downfall of classical mechanics in the microscopic world. The glory of classical mechanics, which was devised in a sophisticated formulation mainly by Lagrange and Hamilton, still provoked some physicists to re-establish the reign of classical physics.

DISCRETENESS AND CONTINUITY

The quantum indeterminacy problem is deeply rooted in the long-standing question of discreteness and continuity. This is an issue which has been under

242 W. Heisenberg, Z. Phys. 43, 172 (1927).

persistent debate since the early days of the Greeks, where we know of Aristotle's critique of the atomism which had been suggested by Democritus, and throughout the Islamic period, which witnessed fierce debates between the philosophers and the Mutakallimūn.

The indeterminacy of quantum states, as described by Heisenberg's uncertainty principle, brought to the attention of physicists the fact that quantum mechanics is the mechanics of an undetermined nature. As noted above, this posed, what came to be known as the "measurement problem" in quantum mechanics. Today, more than three quarters of a century after the advent of the theory, it is still an issue of unprecedented debate. In fact, it is by far the most controversial problem of current research in physics and divides the community of physicists and philosophers of science into numerous opposing schools of thought. The main issues in this division seem to be centred on two things: the quantum jumps and the measurement indeterminacy.

Quantum jumping is an indication of the discrete nature of the atomic world. If this is to be a fundamental characteristic of the microscopic world, then the perceived continuity of the macroscopic world would seem to be illusory. It was reported that Schrödinger once said, "if all this damned quantum jumping were really to stay I shall be sorry I ever got involved with quantum theory."[243] The main difficulty arises when we find differential calculus (which is the backbone of the mathematical formulation of classical physics that was based on continuity and infinite divisibility) to be in need of serious revision. Consequently, the canonical formulations of physical laws would not be valid, and the basic concepts of field theory challenged.

The Schrödinger equation is a deterministic equation and adopts the principle of continuity, as well as the concept of infinite divisibility. However, it is also a wave equation that has helped to provide an approximated picture of the quantum world. The discrete features of the quantum world are now being presented as a product of its wave mechanical nature, which allows for the superposition of waves, thereby generating an interference pattern. Accordingly, one can avoid thinking of the abrupt quantum jumps in favour of thinking in terms of probability distributions, such that some kind of continuity between discrete states is maintained. In this way, instead of having macroscopic continuity becoming an apparent feature, which hides the underlying discreteness, we now have discreteness appearing as an emergent product of some phenomenon of the continuum. In addition to this, it is important to note that precise analysis of the quantum phenomenon of the double-slit interference of particles shows some fundamental characteristic departure from the standard

243 Max Jammer, *The Philosophy of Quantum Mechanics: The Interpretations of Quantum Mechanics in Historical Perspective*, (New York: John Wiley & Sons, 1974): 57.

wave-interference phenomenon.[244] In these experiments a particle continues to be non-divisible; however, at very low intensities, the behaviour of a particle may be shown to be different from the behaviour of photons in an electromagnetic beam. Such a divergence in behaviour awaits an explanation which can precisely identify those features in both of the phenomena that makes them different.

THE APPLICABILITY OF QUANTUM MECHANICS

In this context comes the question of whether quantum mechanics is a theory that can be applied to a single particle, or whether it is a theory of ensembles. Physicists have different opinions on this issue. Some, like Bohr and Heisenberg, believe that quantum mechanics is suitable to describe single particles as well as many particle systems. This is generally the view held by the Copenhagen school. Other physicists, like Einstein and Born, believe that quantum mechanics is applicable to ensembles rather than individual particles and, accordingly, maintain that it can only be interpreted statistically. Still others, such as Everett and Wheeler, believe that quantum mechanics is essentially an interaction theory, which can be realised only through the interaction between the observer and the system. In one way or another, this interaction allows for a subjective interference in determining quantum states. In fact, the basic formulation of the equation of motion in quantum mechanics—Schrödinger's equation—suggests that it can be applied to single particles. On the other hand, having the values of observables coming out as an average may suggest that we are talking about an ensemble of particles, in which each particle enjoys a different value for a given observable. The general behaviour of a system of these particles is then represented by the behaviour of the average. However, this restriction becomes unnecessary if we interpret the existence of an average as occurring due to the many measurements being performed on the same particle. In this case, the implicit fact will be that the value of the observable, which is assigned to the system (the single particle in this case), is not fixed, but is ever changing.

The question arises as to whether this change in the value of the observable is due to the changing state of the system or to the process of the measurement itself. If we assume that it is due to the changing state of the system, then the process of measurement can be taken to be completely passive. On the other hand, if we consider it to be a result of the measurement itself, then we are assuming, primarily, that the measurement itself has a disturbing effect on the system. This amounts to assuming the existence of an interaction between

244 M. Namiki and S. Pascazio, "Quantum Theory of Measurement Based on the Many-Hilbert-Space Approach," *Physics Report* 232, 301 (1993).

the system and the measuring device. Since the microscopic systems under investigation are so small and delicate, no one can deny that such interactions are possible, and may cause subsequent disturbance. Therefore, such interactions will lead to the "decoherence" of the quantum system. The disturbances caused by the measuring devices are generally non-systematic and so complicated that they would be unpredictable. On the other hand, one might expect that, in some cases, the disturbances caused by the macroscopic measuring device would be so large that it would overwhelm the basic value of the observable under measurement.

A third point to be made here is that such disturbances, if known, can be accounted for in the equation of motion through the potential term that is used in that equation. Accordingly, the case will always be that of an interacting system for which the equation of motion may be solved exactly or through numerical techniques. Virtually any environmental factor can be included in the potential of the system, which controls the system's behaviour through the equation of motion. Taking these points into consideration, it would be odd to assume that quantum indeterminacy is simply a result of the incision of measurement.

INTERPRETATIONS OF QUANTUM MEASUREMENTS

In any given individual experiment, the result of the measurement is one of several alternatives. A repetition of the experiment under identical initial conditions may lead to another of these possible alternatives. This is incompatible with the unitary evolution described by Schrödinger. Several solutions have been proposed for this apparent inconsistency. The main ones are:

(1) **The von Neumann Interpretation:** wave function collapse:

To explain the process of measurement, von Neumann suggested that state function can change in one of two different ways:

Process 1: a discontinuous change brought about by the observation by which the quantity with eigenstate $|\psi>$ is projected onto the state $|\phi>=A|\psi>$ instantly, with probability $|<\psi | \phi>|$. This determines the overlap between the state $|\psi>$ and the state $|\phi>=A|\psi>$.

Process 2: a change in the course of time development, according to the deterministic Schrödinger equation.

The description in process 1 is called "the wave function collapse." This means that the state $|\psi \psi >$ after measuring the observable A will be converted into the state $|\phi>=A|\psi>$

A fundamental problem was recognised long ago in this formulation of von Neumann's. This problem embodies the apparent inconsistency between the indeterministic nature of process 1 and the deterministic nature of process 2. This apparent inconsistency has been presented in different forms, and has, in fact, been deeply rooted in the formulation of quantum mechanics from its very beginning.

Josef Jauch[245] presented the problem as follows: the problem of measurement in quantum mechanics concerns the question of whether the laws of quantum mechanics are consistent with the acquisition of data concerning the properties of quantum systems. This consistency problem arises because the system to be measured and the apparatus, which is used for the measurement, are themselves systems which are presumed to obey the laws of quantum mechanics. The evolution of the state of such a system is, therefore, governed by the Schrödinger equation. However, the measuring process exhibits features which are apparently inconsistent with Schrödinger-type evolutions. The typical process ends with the establishment of a permanent and irreversible record, and this contradicts the time-reversible nature of the Schrödinger equation. So, despite the fact that the von Neumann interpretation of quantum measurement was adopted by the Copenhagen school, it, nevertheless, suffers from some fundamental problems.

The statistical interpretation. For this we have two viewpoints:

Viewpoint I, in which quantum mechanics is understood to apply to ensembles and not single particles. Albert Einstein was an advocate of this interpretation, saying, "The function Ψ does not, in any way, describe a condition which could be that of a single system: it relates rather to many systems, to 'an ensemble of systems' in the sense of statistical mechanics."[246] Einstein hoped that in the future a more complete theory might describe quantum mechanics as an approximation of a more general one.

Viewpoint II, proposed by Born and supported by Bohr, according to which the wave function ψ was understood to be a symbolic representation of the system and that $|\psi(x)|^2=|\psi^*(x)\psi(x)|$ is taken to describe the probability density for the system which is in the position *x*. But probability can only be understood to have a meaning through a population. In this case, the population is that of many repeated measurements. This may be asserted by the fact that Born was of the opinion that his suggestion had the same content as that of Einstein's, and that "the difference [in their views] is not essential, but merely a matter of language."[247]

245 J. M. Jauch, "The Problem of Measurement in Quantum Mechanics," in *The Physicist's Conception of Nature*, edited by J. Mehra, (Boston: D. Reidel Publishing Company, 1973), 84.
246 A. Einstein, "Physics and Reality", *Journal of the Franklin Institute* 221, (1936): 349.
247 M. Born, *The Born-Einstein Letters* (Walter and Co., New York; Macmillan, London 1971), 10.

One can say that Einstein's interpretation is covered by the fact that in any measurement on a quantum system we measure macroscopic quantities; a fact which was originally emphasised by Bohr. If, however, we come to measure by any means a microscopic quantity, then the Einstein interpretation will not be valid. On the other hand, by requiring that many measurements are to be made on the same system, Born's interpretation implicitly assumes that the system is to remain within the same state over the duration of all those measurements. Obviously, this cannot be generally guaranteed.

The hidden variables interpretation

This interpretation was championed by David Bohm,[248] who assumed that quantum mechanics was incomplete, and that there were some hidden variables that should complement the physical description in order to get the full picture of the physical world, which is assumed to be deterministic. There are several kinds of hidden variable theories: some of which are local and some non-local. Belinfante[249] has given a very detailed account of these theories, both in their scientific content and in their historical development. According to Bell's theorem,[250] the local hidden variable theories were shown to be inconsistent with quantum mechanics. It may be added that none of the existing non-local theories is found to conclude with any prediction that is new to the standard formulation of quantum mechanics.

The multi-world interpretation

This was originally proposed by Hugh Everett in 1957. Everett reformulated the process of measurements by abandoning the concept of wave function collapse which had been set by process I of the von Neumann's formalism, while keeping the assumption of the deterministic evolution of the system as set out by Schrödinger's equation. Everett criticised the need for "external observers" in order to obtain measurements by the von Neumann scheme, and instead considered the system as being composed of two main subsystems: the object and the measuring device (or observer). This formulation established the concept of "relative state." This treatment led Everett to conclude that:

> Throughout a sequence of observation processes there is only one physical system representing the observer, yet there is no single unique *state* of the observer (which follows from the representations of interacting systems). Nevertheless, there is a representation in terms of a *superposition*, each element of which contains a definite observer state and a corresponding

---

248 Bohm, D. "A Suggested Interpretation of the Quantum Theory in Terms of 'Hidden' Variables, I and II," *Phys. Rev.* 84, (1952): 166.

249 Belinfante F.J., *A Survey of Hidden Variables Theories*, (Oxford: Pergamon Press, 1973).

250 Bell, J.S., "On the Einstein Podolsky Rosen paradox," *Physics 1*, (1964): 195.

> system state. Thus, with each succeeding observation (or interaction), the observer state "branches" into a number of different states. Each branch represents a different outcome of the measurement and the *corresponding* eigenstate for the object-system state. All branches exist simultaneously in the superposition after any given sequence of observations.[251]

Everett further suggested that:

> The trajectory of the memory configuration of an observer performing a sequence of measurements is thus not a linear sequence of memory configurations, but a branching tree, with all possible outcomes existing simultaneously in a final superposition with various coefficients in the mathematical model. In any familiar memory device the branching does not continue indefinitely, but must stop at a point limited by the capacity of the memory.

John Wheeler supported Everett's theory, emphasising its self-consistency,[252] while Graham,[253] working under the supervision of Bryce DeWitt, elaborated on Everett's interpretation. It was assumed that the eigenvalues associated with the observer's subsystem form a continuous spectrum, whereas the eigenvalues associated with the object form a discrete set. In order to reconcile the assumption that the superposition of the object system never collapses—with the ordinary experience which ascribes to the object system (after the measurement) only one definite value of the observable—it was proposed that the world would be splitting into many worlds, all existing simultaneously, with each of them corresponding to a definite component of the superposition. In each separate world, a measurement yields only one result, although this result differs, in general, from one world to another.[254]

Max Jammer goes further and points to the doctrine of re-creation which was suggested by the Mutakallimūn saying:

> Just as in the ancient Muslim school of the Mutakallimūn (*kalām*) the daring assumption of a continual dissolution and recreation of the universe reconciled the apparent continuity of the macroscopic phenomena with the atomic doctrine of space, time and matter, so in the modern theory of

251 H. Everett III, "Relative State Formulation of Quantum Mechanics," *Rev. Mod. Phys.* 29, (1957): 454-62.

252 J. A. Wheeler, "Assessment of Everett's 'Relative State' Formulation of Quantum Theory," *Rev. Mod. Phys.* 29, (1957): 463.

253 N. Graham, The Everett interpretation of quantum mechanics, Ph.D. thesis University of North Carolina at Chapel Hill, 1970; Bryce S. DeWitt and Neill Graham, eds., The Many Worlds Interpretation of Quantum Mechanics: A Fundamental Exposition by Hugh Everett, III, with a Paper by J. A. Wheeler, B. S. DeWitt, L. N. Cooper and D. Van Vechten, and N. Graham (Princeton, New Jersey: Princeton University Press, 1973).

254 Jammer, M. *Philosophy*, 512-13.

> Everett, Wheeler, and DeWitt, or briefly EWD theory, the no less daring hypothesis of a continual splitting of the world into stupendous number of branches reconciled the continuity of microphysical processes with the experience of distinct outcome of measurements. In fact, despite the similarity in the consequences of the principle of re-creation and the EWD theory, one has to recognize the conceptual differences between the two. Whereas the EWD theory is suggesting that the values we are measuring belong to different worlds, the re-creation proposal of *kalām* suggests that it belongs to one and the same world. It raises the question whether we are living in one or multiple world, a mid-blowing question, as this may lead to inconsistency. I find the *kalām* re-creation principle more realistic in describing the states of the different worlds, not as being simultaneously existing but as being different worlds in succession.

However, this topic needs to be discussed further in order to understand how the world remains connected within the valid causal relationships.

THE RE-CREATION POSTULATES

In order to interpret quantum measurements, I propose the following two postulates:

Postulate P (1): All physical observables of a system are subject to continued re-creation.

Postulate P (2): The frequency of re-creating an observable is proportional to the total energy of the system; a system may be a whole body or an individual particle. It will be shown below that the re-created observable assumes a new value each time it is re-created. This will cause the observable to have a distribution of values over a certain range (width) that is always controlled by the re-creation frequency. The higher the total energy of the system, the narrower is the range of values over which the dispersion is expected to occur (and vice versa). For this reason, macroscopic systems are expected to behave classically, whereas microscopic systems exhibit mostly quantum behaviour. Clearly, the narrower the dispersion of values, the more determinable is the value of the observable, and vice versa.

RE-CREATION AND THE UNCERTAINTY PRINCIPLE

Once created, an observable assumes a given base value, as defined by the state of the system at that moment. Re-creation is a process of change. According to the re-creation postulate P (1), physical observables are in a natural process of continued re-creation, irrespective of the measurement operation. However, the values of these observables can only be known at the time of measurement. Once a given observable is re-created, the other observables of the system will also be affected, thus changing their values in accordance with the related physical laws. Any change is best described, in the most general form, by the generator that corresponds to that respective observable. For example, if x is re-created then the system will change infinitesimally by $\partial/\partial x$; this, however, is simply proportional to the momentum operator. This will duly cause the value of position x to change every time it is re-created, thus presenting a distribution of values for x instead of one single value. Conversely, if momentum p is re-created, then the whole system will change by $\partial/\partial x$, but this will cause a shift in the value of position x, since the momentum is a generator of translation according to quantum mechanics. Therefore, every time x is re-created, a change in the momentum of the system will occur; that is to say, upon re-creating the position, a momentum is created. Conversely, every time momentum is re-created, a change in the value of the position will occur. This means that re-creating the position will result in creating momentum, and vice versa. If the system itself is to stay invariant under the process of re-creation, then we must have: $\left(\frac{\partial}{\partial x}x - x\frac{\partial}{\partial x}\right)|\psi\rangle = |\psi\rangle$

By using the explicit forms for both position and momentum operators, this would imply that $p\,x - x\,p = [\,p, x\,] = i\hbar$, where x p are the position and the momentum operators, respectively. In other words, the effect of change is logically seen as a commutation of the operator and its generator (which are also known as complementary observables). This well-known commutation relationship led to Heisenberg's uncertainty relations. In this scheme, however, measurements could be passive actions that do not necessarily affect the system itself.

This proposal of re-creation preserves the statistical nature of the possible values of the observables and resolves the question of whether quantum mechanics is applicable to single particles or to an ensemble of particles. Here, we see that as the single particle state gets continually re-created, an ensemble of values for the same physical observable is formed. This ensemble of values could be subjected, of course, to statistical analysis. Practically, measurements of an observable, taken over a duration of time that exceeds the re-creation period, will always yield an average of the values assumed by the system during that period of measurement. So, in practice we measure average values of an ensemble of instantaneous values every time we perform a measurement. This explains how probabilistic behaviour arises in the case of a single particle quantum system. According to the above scheme, we always measure average values with very low dispersion for macroscopic objects; the re-creation frequency is very high and, consequently, the measurement time cannot cope with the re-creation period. This gives the macroscopic world its classical, and what appears to be deterministic, characteristics. This is why the measured values of the observables of a macroscopic system are always very close, possibly even identical, to the theoretical expectation values of the observables. On the other hand, in microscopic systems, the re-creation frequency is relatively low and, therefore, we would expect the dispersion of values to be high enough to reveal the indeterministic character of the world.

This proposal also provides us with a better understanding of the origin of the uncertainty relations. Here, we see that the appearance of uncertainty in the values of complementary observables is a direct result of re-creation, and the entanglement of such variables. This means that indeterminism is a direct consequence of continued re-creation.

PHYSICAL IMPLICATIONS OF RE-CREATION

There are several implications of the proposed re-creation scheme described above, some of which can be used to test the theory. However, because of the largely technical nature of these implications, I will only provide an overview of those which may be of interest to people working on these issues in the science and religion debate. The full technical treatment of these implications will need to be presented elsewhere.

Macroscopic Quantum States

The re-creation frequency can be affected by an external force field. Since it is known from the Theory of General Relativity that the duration of an event occurring near a gravitational field is dilated by a factor proportional to the strength of the field, then one should expect re-creation periods to be dilated once they are in the vicinity of a strong source of gravity.[255] Consequently, re-creation frequencies should be red shifted when they are in the vicinity of a strong gravitational source, which means that macroscopic classical processes would start to exhibit quantum features when they are in a strong gravitational field. This will cause the appearance of macroscopic quantum states in such regions, for example, near the event horizon of black holes.

Quantum Coherence

Coherence is a phenomenon in nature whereby the efficient transformation of energy takes place. In the microscopic world, coherence occurs when two oscillating systems resonate with the same frequency and are in phase; that is, a system comprising two adjacent pendulums. However, we do not usually talk about such resonating systems in the microscopic world as being coherent states. This phenomenon is best realised in quantum systems, which always display high efficiency in energy transformation, for example, lasers.

255 See for example: S. Weinberg, *Gravitation and Cosmology*, (New York: John Wiley & Sons, 1972).

The availability of a macroscopic quantum state may make it plausible to expect the occurrence of macroscopic coherent states too, thus opening the way to understanding some very obscure phenomena (such as gamma-ray bursts), which are known to occur at the far rim of the universe. In addition to this, the re-creation postulate (P1) allows for a new definition of coherence, by which two systems can be considered in a coherent state if their re-creation frequencies are identical and their re-creation happens to be in-phase.

Quantum Zeno Effect

This interesting proposal was suggested by Misra and Sundarshan[256] in 1977. The proposal is based on the notion of wave function collapse, and was considered to be a prediction of the collapse interpretation. The idea is that, if continuous measurements are carried out on a given state, then the system is expected to stay in that state because of the continuous collapse of the wave function into the same state (as they say, a watched pot never boils!). Although this prediction was independently verified by Itano *et al.*[257] it was subsequently refuted by Petrosky *et al.* .[258] Recently, rigorous calculations have been made which present the quantum Zeno effect (QZE) quantitatively in a more accurate form by taking into consideration the effect of the measurement duration.[259]

The re-creation interpretation presented in this chapter sets an upper limit for the measurement time for the QZE to be verified. The measurement time of an observable (e.g. transition energy) should be less than the re-creation period needed for the QZE to occur. Measurements performed within a time duration, which is more than the re-creation time, will result in averaging the values of the observable over several re-created states, and, consequently, cannot hold the system in a specific state. Accordingly, QZE will not be verifiable if the measurement time is more than the re-creation time.

Quantum Entanglement

This is a very interesting phenomenon which was predicted by Albert Einstein and his colleagues[260] in a proposed thought experiment. Pairs of quantum systems originating from the same state, say an electron-positron pair created from the same gamma-ray photon, are shown to exhibit non-causal entanglement. For example, if the direction of the spin of the electron is changed, then the direction of spin of the positron will immediately change to comply with

256 B. Misra and E.C.G. Sudarshan, J. Math. Phys. 18, (1977): 756.

257 W.H. Itano, D.J. Heinzen, J.J. Bollinger and D.J. Wineland, Phys. Rev. A 41, (1990): 2295.

258 T. Petrosky, S. Tasaki and I. Prigogine, Phys. Lett. A 151, (1990): 109.

259 L.S. Schulman, Physica Scripta 49, (1994): 536-542; J. Phys. A: Math. Gen. 30, (1997): L.293-299.

260 Einstein, Albert, B. Podolsky, and N. Rosen. "Can Quantum-Mechanical Description of Physical Reality Be Considered Complete?" *Physical Review* 47 (1935): 777–80.

the new state of the electron. This is independent of the distance between them. Einstein called this a "spooky action-at-a-distance", which violates the local causality of the Theory of Special Relativity. Accordingly, he rejected this result and concluded that the theory of quantum mechanics was incomplete.

Despite the refusal of Einstein to accept the possibility of such an event occurring, the phenomenon was verified experimentally by Alain Aspect[261] in 1982. No satisfactory explanation for this quantum entanglement has been put forward until now. However, the re-creation mechanism might explain this phenomenon as particles created from the same initial state ought to have their re-creation process interconnected so as to comply with the conservation requirements of the original system. Such a connection would appear to be established during the phase of re-creation. A single state being split into two states will generate two states with frequencies that are re-created in phase, even though the two frequencies might be different. This phase correlation between the two separated states may explain the apparently non-causal connection between them. Entanglement is a requirement of the phase conservation.

DISCUSSION AND CONCLUSIONS

The scheme proposed in this chapter for the interpretation of quantum indeterminism offers a scope that allows for an objective ontology of the physical world, in addition to the possibility of being undetermined. Such a scheme is more realistic and more consistent than the observer-dependent analyses which are implied by the von Neumann and the Everett-Wheeler interpretations.

Unlike Schrödinger's cat and the EPR paradox, the re-creation scheme is free from the known paradoxes of quantum measurements since it does not consider a subjective role for measurements or a wave function collapse, while assuming the natural presence of the entanglement of states belonging to the same system. Moreover, this scheme resolves the statistical nature of quantum mechanics by allowing the statistical distribution of the possible values that an observable might hold so as to fall within the natural process of continued re-creation of that observable.

It is important to note that the above scheme will not affect the standard calculations of quantum mechanics, except that it might motivate new investigations into regions which, until now, have not been excavated by mainstream research. These could include the existence of macroscopic quantum states and the possibility of understanding the gamma ray burst as being the result

261 Aspect, Alain, Philippe Grangier, and Gérard Roger. "Experimental realization of Einstein-Podolsky-Rosen-Bohm Gedanken experiment: a new violation of Bell's inequalities." Physical review letters 49, no. 2 (1982): 91.

of macroscopic quantum processes taking place under very specific conditions deep within the universe.[262] However, the scheme proposed here is by no means complete and is, therefore, open to further development.

262 Gehrels, Neil, Luigi Piro, and Peter J.T. Leonard. "The Brightest Explosions in the Universe." *Scientific American* 287, no. 6 (2002): 84-91.

Chapter Nine

Motion

The fact that there was much controversy about the concept of motion in *kalām* still has a special flavour because it contains rich elements that make it a composite concept, unlike our common, simple understanding of motion in terms of covering a given distance within a duration of time. Philosophically, motion was understood as a change; that is, any change can be considered a kind of motion. Philosophers and scientists have expressed their views about motion in various ways, but motion as transport in space (*naqla*), had a special meaning for the Mutakallimūn, as it was one of their prime arguments in demonstrating that the world is temporal, as I have mentioned before.

In this chapter, I will concentrate on the concept of motion as understood by the Mutakallimūn, without referring to its special importance historically. We should note primarily that motion in *kalām*, as with other philosophical trends, was treated only qualitatively and that no serious quantitative study of it is found with any of the Mutakallimūn. Such quantitative studies are found with those scientists dealing with physical motion, such as Ibn al-Haytham (965–1040), Abū Al-Hasan al-ʿAmirī (d. 992), al-Bīrunī (973–1048) Ibn Malkā (1080–1164), and many others.

DEFINITION OF MOTION

Al-Jurjānī, in his lexicon, defines motion in space as "the exposition from what is in force to what is in action gradually."[263] He further states that *motion*, "is said to be the occupation of a place after being in another", whilst pointing out that "it is also said that motion is two beings (*kaūnayn*) at two times in two places, and *rest* is two beings in two times in one place."[264] However, Al-Ashʿari refers to Abū al-Hudhayl al-ʿAllaf, and adds that when Abū al-Hudhayl talks about two beings existing at two different times and in two different places, he means that the object has been in one place at one time and in another place at another time. Therefore, to be at rest is to be in the same place at two successive moments of time. This has been asserted by al-Qalānisī, as reported by Al-Ashʿari .[265]

MOTION ACCORDING TO THE MUTAKALLIMŪN

The motion of a particle is visualised, according to the Mutakallimūn, as a succession of states of being describing the existence of the particle at different positions. As the particle is being re-created so is its spatial occupation (the position at which it exists) and its temporal moment of existence. Al-Ghazālī gives an elegant description of motion according to *kalām*, saying, "the states that follow each other through continuous periods of time are described as movements only because they alternate by continuously originating anew and continuously ceasing to exist… the essence of motion is inconceivable without also conceiving nonexistence to follow existence."[266] This means that a particle in motion does not translate through pre-existing space. Instead, it actually occupies successive positions in which it appears suddenly. This exposes the difference between the understanding proposed by al-Naẓẓām; i.e. the jump, or leap (the concept of *ṭafra*), and the rest of the Mutakallimūn. Whereas al-Naẓẓām assumes that a space exists between the successive positions of the particle in motion, the rest of the Mutakallimūn did not accept this and assumed that the occupancy of the position occurs upon re-creation.

Ibrahim al-Naẓẓām considered motion to be perpetual, meaning that no object can ever be at rest. He is reported to have said, "I don't know what rest is except that an object being in the same place at two successive moments, meaning that it has moved within it at two moments of time."[267] This understanding of al-Naẓẓām suggests that an object can never be at absolute rest, but

263 Al-Jurjānī, al-Tarīfāt: see entry, al-Haraka.

264 Ibid

265 Al-Ashʿari, *Maqālāt* 324.

266 Al-Ghazālī, Moderation in Belief, Aladdin M. Yaqub, Chicago University Press (2013), 44.

267 Ibid. 325.

could be stationary only (stationary means that a body is in motion without covering a net distance). This means that such an object should make two successive moves in opposite directions, an idea of al-Naẓẓām's which echoes in modern times with the problems of the quantum mechanical simple harmonic oscillator and of a free particle in a rigid box, where we find that the particle minimum energy is always more than zero.

The *Ṭafra* (Leap) of Al-Naẓẓām

The above is not the only case where al-Naẓẓām seems to have had a vision of quantum mechanics. He is reported, by many authors, to have said that a particle in motion makes successive jumps, each leap being the distance between two positions. This leap he called *ṭafra*. Al-Naẓẓām found that the leap of motion is necessary to explain how the body can cover an infinitely divisible distance in finite time. With this suggestion, we have a resolution for the well-known Zeno paradox, which was presented as an argument against infinite divisibility of the finite distance between two points.

Al-Naẓẓām is further reported to have said that the body is, "the long, wide and deep with no finiteness for its part, as there can be no half unless it can be halved and no part except what can be parted."[268] This means that a finite distance can be subdivided into an infinite number of parts. Consequently, and in order to explain how motion occurs and how such an infinite number of parts of the distance can be covered in finite time, al-Naẓẓām had to assume his genius leap. Al-Ash'ari reports on this, saying that, "people have different views about the leap (*ṭafra*); al-Naẓẓām claimed that the body could be in one place and move to a third place without passing through the second."[269]

The contemporary philosopher and historian of science Max Jammer commented on this notion of al-Naẓẓām's, saying, "in fact, al-Naẓẓām's notion of leap, his designation of an un-analyzable inter-phenomenon, may be regarded as an early forerunner of Bohr's conception of quantum jumps."[270] This shows that, according to al-Naẓẓām, motion is quantised. This situation is seen in atomic physics where the transition from one orbit, in Bohr's atom, to another can only take place in jumps (a close look at the distances between the orbits in Bohr's model and a simple calculation shows that the electron covers the transition between two levels in about 10^{-22}s. Should the electron stay much longer, it would radiate its energy and fall into the nucleus in 10^{-11} seconds).

268 Al-Ash'ari, *Maqālāt* 304.

269 Ibid. 321.

270 Jammer, Philosophy, 259.

Views of Mutakallimūn about Ṭafra

Most of the Muʿtazilis considered the notion of leap as absurd. Al-Baghdādī reported that, "most of the Muʿtazilis are in agreement in considering al-Naẓẓām an infidel, among these is Abū al-Hudhayl, in a book about accidents and man and the non-divisible part."[271]

Ibn Matawayh, who was a student of ʿAbd al-Jabbār, discussed the notion of *ṭafra* in his book, *al-Tadhkirah*, where he warns at the beginning that the notion of al-Naẓẓām's means, "you can be in Basrah at one time and then you may be in China some time later without passing through the places in between."[272] This sounds like, what we call nowadays, quantum teleportation. Ibn Matawayh, more seriously, suggests that if the notion of *ṭafra* would apply to light rays then, "one would be able to see through the wall as the light ray will jump through it and the prisoner cannot be prevented from leaping away."[273] This kind of conjecture shows that, although some people have really understood the deep implication of the notion of *ṭafra*, they cannot accept such bizarre implications because it does not conform with the prevailing logic. This reminds me of the quantum tunnelling effect, which is exactly what Ibn Matawayh is describing here. Indeed, it is a very advanced vision that led him to comprehend the implications of the quantum leap, as suggested by al-Naẓẓām. This enforces the consistency of the claims presented in this book regarding the contemporary implications of some topics of *Daqīq* al-*kalām*.

Al-Naẓẓām and the other advocates of *ṭafra* came forward with an argument that posed a paradox. They said that the motion of the millstone could not be understood unless we assume that the leap is taking place in the outer parts of the stone, as this outer part should cover the same distance that is traversed by the parts of the stone which are near the centre.[274] The rest of the Mutakallimūn replied by bringing in their concept of stationary points along the path of motion. They argued that the outer parts of the millstone are experiencing less rest points than the parts near the centre.[275] Obviously, this understanding of motion presumes that time and space are discrete and that the motion of an atom or a body through space, along a given trajectory, involves some rest points. The particle travelling along a certain trajectory will encounter moments of rest and moments of motion. A trajectory with a large number of rest points means that the particles will be slower than those taking

271 Al-Baghdādī, (1977). Abdul Qahir, Kitab al-Farq bayn al-Firaq, Beirut: Dar al-afaq al-Jadida.

272 Ibn Matawayh, al-Tadhkirah, 196.

273 Ibid. 198.

274 Ibn Matawayh, al-Tadhkirah, 198.

275 Ibid. 199.

another trajectory which has a fewer number of rest points. It seems to me that both al-Naẓẓām and the Mutakallimūn were basically accepting this motion of the rest points, although they have different views in respect of the divisibility of parts. Ibn Matawayh presents many examples from the debate about the *ṭafra* and continuity and discreteness of motion.

The later Mutakallimūn, who appeared after 1300A.D., held the same view about motion being discrete. It seems that this idea originated from the concept of discreteness of space and time. Al-Rāzī says, "motion and time, each is composed of successive parts that are indivisible."[276] For this reason, motion was also thought to take place in successive steps. Al-Āmidī says, "that slowness in motion is caused by the presence of many rest points, while fastness is due to its less number."[277] Al-Ījī expresses the same understanding that a slow body has more rest points along its trajectory whereas a fast body has less of them.[278]

Analysing the *kalām* views about motion, we can say that it is viewed as successive moments of unrest along a trajectory which is filled with rest moments at which the moving body is thought to be still. A slow body is that which follows a trajectory containing many rest points while a fast body is the one which is following a trajectory with fewer number of rest points. This implies that there is one universal speed and several different trajectories. So, here we have two concepts of motion; the *kalām* concept, which assumes one universal speed but different trajectories, and the philosophical concept, which assumes one trajectory with different speeds. However, one may think that the *kalām* concept could face difficulties in explaining the motion of two bodies on the same trajectory: one fast and the other slow. Such a case cannot be explained by saying that for each body there is a specific trajectory, and that no two particles can be on the same trajectory with different speeds, since the trajectory of a body (or particle) has a path related to the particle itself and has nothing to do with space. This means that the trajectory is not a prescribed path that the moving particle has to follow but it is the path generated by the motion of the particle, instantaneously. This is a novel understanding of motion that may have some interesting implications, particularly when it comes to stating that particles acquire their mass as they move through the Higgs field.

The notion of rest points along the trajectory of a particle may be related to the state of re-creation of the particle. A particle with a high rate of re-creations is expected to experience a greater number of rest points, while a particle with a low rate of re-creations may experience fewer rest points. The point I am making here, is that the number of rest points is directly proportional to the number of

276 Fakr al-Rāzī, Usul 35.
277 Āmidī, Mubīn, 95-96.
278 Al-Ījī, Mawāqif, 21272-280.

pulses per unit of time that a particle may represent. Such ideas may have some useful implications that could help to understand the motion of elementary particles in free space.

FORCE AND ACCELERATION

One can hardly see a clear conception of acceleration with the Mutakallimūn, nor can one see the difference between the concepts of speeding and acceleration in *kalām* discussions. However, one finds that there are some views that were expressed in relation to freefall and what happens to a stone that is thrown upward or pushed in a certain direction. This topic has been mostly discussed in the context of generation (*tawlīd*) and adherence (*Iʿtimād*) by the Muʿtazilis. These discussions, in fact, are closely related to the subject of causality, and I will return to this point when discussing causality in the next chapter.

However, it might be interesting to know that the Mutakallimūn have discussed the forces effecting the motion of a stone thrown upward, where they realised that the stone reaches a height at which it stops and then falls down. The Muʿtazilis explained this by saying that the stone thrown upward rises by the forced adherence (*I ʿtimādāt mujtalaba*) and will come down by the necessary adherence (*Iʿtimādāt lāzima*). However, they had different opinions about the state of the stone at its highest point, as some of them claimed that the stone would be stationary momentarily, whilst others, like al-Kaʿbi, negated such a possibility.[279] The point of interest to mention here is that none of the Mutakallimūn presented a discussion or argument which attributes the forces on the stone to the gravity of the Earth. It is as if they could not recognise that the Earth attracts the objects above it. For example, al-Jūwaynī reports that ʿAbdul al-Jabbār suggested that a stone thrown upward will fall down because, "the air disturbances in front of the stone reduces its forced adherences so that it gets overcome by the necessary adherences, causing a reversing of its motion."[280] However, al-Jūwaynī seems to have figured out the state of motion of the stone simply by suggesting that it is controlled by the balancing of the forced and the necessary adherences, and that when such a balance occurs, the stone will stop and after that start to fall with acceleration. This is a point that needs to be studied in order to see why the Mutakallimūn did not consider the philosophical viewpoint which considers the Earth as the centre of attraction of all the world; a view which they could have borrowed from the Greek philosophers.

279 Al-Naysābūrī, Masa'il, 195.
280 Al-Jūwaynī, al-Shāmil, 507.

The Weight

This is a fine concept that has been rarely discussed in *Kalām*. However, al-Jūwaynī and al-Bāqillānī point to it in their books, *al-Shāmil* and *Tamhīd,* respectively. It is also discussed in Matawayh's *Tadhkira* and in Ijī's *Mawaqif.* The question, put by the Mutakallimūn concerning weight, asks whether it is an intrinsic property of the *jawhar* or some added entity (emergent property). If it is an emergent property, then it would be an transient (*a'rād*) but if it is an intrinsic property, then it would be a character for the *jawhar.* Al-Jūwaynī says that the Mutakallimūn differed on this question. Some, he says, have considered the weight as an intrinsic property and others have considered it an emergent property. He also points to the position of some Ashʿaris, where he admits that they have adopted the notion of adherence (*Iʿtimād*), with al-Bāqillānī considering weight as a sort of downward adherence of the body.[281] Furthermore, al-Jūwaynī expresses the position of the Ashʿaris towards the notion of adherence, saying that they have accepted it with a fundamental reservation by which they constrain it with the principle of re-creation. He says, "if we are to adopt the adherences, we would not separate them into forced ones and necessary ones, but would say that all are non-enduring and it follows successively in the same way as the accidents."[282] This reflects the fear of the Ashʿaris that adopting the adherences may lead to some sort of naturally intrinsic self-action due to natural objects. This has hindered the Ashʿari approach to studying natural phenomena and has lifted their quest substantially.

The Mutakallimūn discussed many problems concerning floating bodies and the action of different forces on bodies at rest and in motion; many of these discussions are found with the Muʿtazilis. Nonetheless, it remains to say that their treatment was somewhat primitive and does not suggest anything innovative. Such problems are to be referred to the works of the scientists as it deals mostly with technical questions and not philosophical interpretations.

AL-JUBBĀIE ON FREEFALL

Some of the Mutakallimūn believed that a heavier body would fall toward the Earth faster than a light one. This complied with Aristotle's view about freefalling bodies. Other Mutakallimūn believed that it is air which causes the drag of the motion of freely falling bodies. They said that if air is removed, then all bodies would fall towards the Earth at the same speed. Ibn Matawayh tells us that Abū Ali al-Jubbāie held this opinion and argued in defence of it, saying that, "the reason for the difference between the descendance of the

281 Ibid. 490-493.

282 Ibid. 497.

light and the heavy bodies is the air in the atmosphere, otherwise if we drop a stone and a feather together, then they will fall together, but it is the air which prevents the light body from descending [with the heavy] while the heavy body pierces it [the air]." This view of freefall was put forward more than 600 years before Galileo declared the same position and rebutted Aristotle's belief.

In fact, the above understanding of freefall held by al-Jubbāie was well-founded and, by all means, it is a novel understanding of the freefall of bodies that preceded Galileo Galilei by more than six centuries. Although, we don't have a detailed account of this issue in literature from the time of al-Jubbāie, we find Hibatullah Ibn Malka al-Baghdādī (1080–1164) in his book, *Kitāb al-Mu'tabar* fi *al-Hikmah* (The Considerate Wisdom), explaining, in some detail, the physical basis of this view of the Mutakallimūn, without naming any of them. He claims that they arrived at such a conclusion as they assumed the existence of a void. Aristotle, in his opinion, was not able to reach such a conclusion because of the constraints imposed by his basic assumptions about motion; specifically, his assumption that space is full, thereby negating the existence of voids. More investigations on this issue are needed in order to clarify the contribution of *kalām* on such fundamental topics in Islamic physics.

MOTION IN THE NEO-KALĀM

The principle of atomism suggests that motion should be discrete, and the principle of re-creation would certainly support such a proposal. As we have seen in the previous chapter, the new vision of re-creation provides us with all the known aspects of quantum systems. This could provoke one to suggest an explanation of motion on the microscopic level, particularly in quantum systems.

The problem of particle interference when passing through two adjacent slits is well known. In 1965, Richard Feynman presented a thought experiment to show how this happens; this is well described in his lectures.[283] Many physicists, including Feynman himself, expressed their astonishment at this behaviour of particles, and the conclusion was reached that the particle appears to be passing through the two slits simultaneously; otherwise, the result of the experiment cannot be explained.

Recently,[284] an experiment was conducted where single electrons were sent successively through the two slits. It was shown that the particle distribution

283 Feynman, R, P., Leighton, R.B., and Matthew, S. (1965). *The Feynman Lectures on Physics*, vol. 1, Reading, MA: Addison-Wesley., chap1.

284 Bach, R. et al., Controlled double slit electron diffraction, *New Physics Journal*, 15, 033018 (2013).

on the screen made the same interference pattern that it made when a beam of particles used: see Fig (1) below.

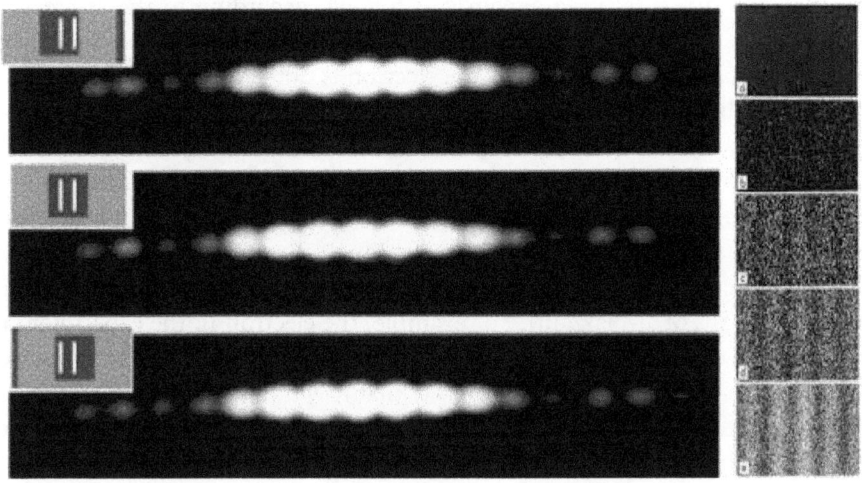

Fig (1) The pattern of the distribution of particles after a long time of bombardment.

This might indicate that the particles were directed to their positions on the screen by some sort of mysterious guiding agency. Perhaps this result might bring to the front Bohm's explanation, which suggests a "guiding wave"[285] explanation for this experiment, which I admit is an intriguing one. However, we have to remember that Bohm's interpretation of quantum mechanics was rejected by the majority of quantum physicists for many reasons, and the local hidden variables theory has been refuted by the experimental verification of Bell's theorem.

Fig. (2) Caricature expressing the astonishment of the physicists from the results of electron double slit diffraction.

285 Bohm, D. and B.J. Hiley, *The Undivided Universe*, 32-33.

I find that the re-creation principle can offer a solution here, and can explain why the trajectory of an electron looks as though the electron has passed through the two slits simultaneously as follows, according to the re-creation formula, the electron re-creation rate is about 8.0×10^{21} Hz. In the experiment above, the kinetic energy of the electron is 600 eV, which means that its velocity is about 1.45×10^{7}m/s. The electron then covers a distance of 1.8×10^{-15} m between the two successive re-creations. The distance between the two slits, as reported in the paper, is 272 nm. This means that the electron will get re-created more than 1.51×10^{8} times between the two slits; this rate allows the electron to appear as if it has passed through the two slits simultaneously.

The re-creation proposal enables us to calculate the number of rest moments (or leaps of al-Naẓẓām). For example, the electron in the experiment of the two slits described above will cover a distance of 1.8×10^{-15}m between two successive re-creations. Thus, we can say that it will have about 5.5×10^{14} rest points (or leaps) per metre. The faster the particle, the fewer the number of rest points (leaps) per metre, and the slower the particle the greater the number of rest points along its trajectory. This description provides us with a consistent explanation of motion, in accordance with the principle of atomism and the principle of re-creation. After all, al-Naẓẓām's leaps and path atomism are complementary.

Chapter Ten

Natural Causality

Causality is one of the most important subjects in *kalām*. This topic is fundamentally related to the basis of Islamic belief; that is the acknowledgement of the existence of Allah and his dominance over everything in the world. No wonder then, that this topic is considered of the utmost importance when it comes to assessing the belief of Muslims, and the degree of their submittal to the oneness of Allah. This is the concept of *Tawhīd.*

A cause is understood to be an element that is identified to be responsible for producing an effect whenever all related conditions are satisfied. There are two types of causality: natural causality, which is related to those causal elements associated with natural beings and the innate properties of objects, as we see in the process of fire burning cotton or medicine curing illness, and human causality, which is related to causal elements associated with the wilful acts of humans. Natural causality is a topic of *Daqīq* al-*kalām*, while human causality is a topic of *Jalīl al-kalām*.

The study of causality has gained importance through being a basic concept for understanding the relations between the parameters that contribute to any natural phenomenon. It is the most basic expression for the relation between those factors which are at play in the occurrence of events taking place in our world. To understand the causes means to identify the essential parameters and conditions that contribute to the occurrence of the phenomenon.

In the analysis of many of the phenomena taking place in the world—for example, the freefall of a stone towards the centre of the Earth, the ignition of a material as it nears a source of heat, or the recovery from illness after taking medicine—the events indicate that those essential parameters and conditions cause the occurrence of the related effects. So, here we have to identify two levels at which causality must be comprehended. The first is to identify the essential parameters that help bring about the phenomenon as being the cause of the related effects, and the second is to attribute the observed effects in the phenomenon to those parameters and conditions (causes) uniquely; namely, to consider them the actual creators of the phenomenon.

On the first level, we need not assume any mindful prevalence throughout, except what might be necessary for the organisation of the process, but on the second level we have to ascribe some mindful control embodied in those causes and a purpose for the phenomenon (the effect). It is not sufficient to claim that it is the nature of the cause that leads to the related effect, unless the term "nature" is fully understood, as well as the mechanism by which the specific cause produces the related effect unambiguously.

In this chapter, I will investigate the concept of natural causality from an Islamic perspective, more specifically in the light of the propositions of *Daqīq* al-*kalām*, as discussed in a scientific context. This topic has become one of the hot issues that needs to be resolved in order to have a proper understanding of the relationship between science and Islam. The topic is one of the major applications of *Daqīq* al-*kalām* and, as we will see, a new insight would provide us with a better vision of the subject other than the one that has been commonly spread and understood by many people. The aim of this chapter is to present one of the fundamental problems of the new *Daqīq* al-*kalām* as an example which can be further pursued along the same lines.

I make no attempt to consider either the historical development of the problem of causality or the views of philosophers and thinkers, other than those Muslim scholars who were concerned with the subject, since such matters can already be found in other works.[286] Through the work presented here, I will try to provide an Islamic view which I believe is in agreement with the stipulations of the Qur'an, and I will try to recast those stipulations in a modern scientific and philosophical context that subscribes to the main principles of *Daqīq* al-*kalām*. Therefore, I can say that this chapter is a major exercise of this trend that shows how we can reconcile our belief system with our scientific knowledge. The goal here is to demonstrate how the question of causality can be more accurately understood, so as to comply with what we have learned over the many ages of investigating the world.

---

286 See for example, Wolfson, *kalām*, 518-600.

SOME BASIC DEFINITIONS AND CONCEPTS

There are a number of terms that are related to the concept of causality, and which can be defined in slightly different ways using different terminologies. Below, I will present some basic definitions and concepts relating to causality and causal determinism. Since these terms are mostly used in their philosophical context, it is expected that there will be different expressions for them embedded within different philosophical views. I will define terms as they are to be used in the text, in order to avoid confusion. For this reason, no specific references for these definitions are given.

Cause (*sabab*) and **Reason** (*ʿilla*) are commonly used by most of the jurisprudents (*fuqahāʾ*) of Islam synonymously. Linguistically, the two terms are different, and differentiating between them might be useful in designating the proper explanation and attribution of factors at play in any event. In Arabic, the *cause* (*sabab*) is said to be the means through which either an event might take place, or an aim achieved. In *Lisan al-Arab*, composed by Ibn Manẓūr, we read that, "the cause is everything by which you get to another thing. The plural is *asbāb*."[287] In most of the Arabic original use, the cause is a means by which a certain target is achieved. This is why they call a piece of rope the cause (*sabab*), because with it you can reach a place or pull something to get it. Reason (*ʿilla*), in Arabic, is meant to be justification or the explanation that can be given for the occurrence of an event or a decision taken. This is why disease is called (*ʿilla*), since it explains the illness.

Causal Relationship: A causal relationship is an expression describing the relation between a set of factors or entities (causes) that are found to be present in conjunction with a phenomenon (the effect).

Nature: This is a very important term that will be used extensively in this chapter and, therefore, it is important to establish its definition. I use the term "nature" to mean either of the following:

(1) the innate properties of things (in Arabic, *ṭabʿ*)

(2) phenomena observed in the physical world, or the whole of the physical world itself. Misplacement of these two meanings for the term "nature" may result in much confusion and a distortion of the intended meaning.

Predictability: This term refers to the expectation of an outcome or an event. The prediction should be based on certain logical consequences which are set according to certain knowledge, otherwise it will be mere unfounded prophecy. Predictability is directly related to expectation. Expectations are made according to certain experiences. Experience needs memory but not necessarily comprehension, implying that there should be history.

287 Ibn Manẓūr, *Lisan*, 1/458.

Predictability is one of the most desired features of modern scientific theory, since it is a good test of a theory. Predictability does not entail determinism because there are highly predictable systems that are indeterministic. For example, quantum systems are highly predictable although indeterminate, whereas chaotic systems are completely unpredictable while being fully deterministic.

Determinism: In the *Stanford Encyclopedia of Philosophy*, it is said that, "the world is governed by (or is under the sway of) determinism, if and only if, given a specified way things are at a time, the way things go thereafter is fixed as a matter of natural law."[288] This description of the term "determinism" requires a law that determines occurrence. This alleged law is called "*natural*" and here we face the problem of whether it is a law of nature that specifies deterministic behaviour or whether it is something else. We can also question whether such a "natural" law exists or not. These questions will be discussed below. Nevertheless, I feel that the above description is accurate enough to describe the concept of determinism. In fact, to a great extent, the macroscopic world that we deal with in everyday life is deterministic.

Causal Determinism: Again, in the *Stanford Encyclopedia of Philosophy*, we read that this concept points to, "the idea that every event is necessitated by antecedent events and conditions, together with the laws of nature." Here, we can question whether such an association between events is such a necessity that without it, nature could not function, and whether such an association fully determines the causal relationships. The laws of physics describe such relationships, and specify the variables and invariables involved in natural phenomena. This is why the laws of physics are deterministic, even though they might be describing indeterministic phenomena. For example, the Schrödinger equation describes deterministic causal relations while the related phenomena (solutions to the equation) include entities that obey probabilistic distribution.

THE PHENOMENOLOGY OF CAUSE

Determinism means that an event described by a physical law must occur once the set of necessary and sufficient conditions is available. For example, the conditions for two bodies to attract each other, according to Newton's theory of gravity, require that both should have a mass and that they should be separated by a finite distance. However, these conditions are not necessary in Einstein's theory of gravity, since a massless object can feel gravity as it passes by a massive object. The difference between the two theories lies in describing the force of gravity as either a field emanating from a mass, according to Newton, or a spacetime curvature, according to Einstein. The sufficient condition for a

288 Carl Hoefer, "Causal Determinism", in *The Stanford Encyclopedia of Philosophy*, ed. Edward N. Zalta, Stanford University, Winter 2008.

cause is something else. It is not an easy task to define the sufficient conditions for any cause, since there may always be something unknown which contributes to that set of conditions.

I have mentioned that the necessary condition for two bodies to attract each other is that, according to the Newtonian picture of gravity, they must have individual masses. However, this is not a sufficient condition, even in Newton's picture. We need to add to this condition that the two bodies should be electrically and magnetically neutral. For example, if the bodies are charged with similar electrical charges, then they will repel each other, and their attraction will be affected. Similarly, if the two bodies are magnetically charged, then magnetic forces will affect their gravitational attraction too. In this case, Newton's gravity cannot be properly verified. There may always be other conditions that will prove to be effective on a large or small scale. The Theory of Special Relativity has shown us that Newtonian mechanics ignores some effects that are only observable at high speeds, whilst the Theory of General Relativity shows us that there are also certain effects that are ignored by Newton's law of gravitation; these effects become very important in strong gravitational fields. In addition, it is known that quantum physics revealed important facts about the microscopic world that were not described by classical physics. The Theory of Relativity and quantum physics have drastically changed the concepts and mathematical formulations of classical physics. New theories imply new conditions and new concepts suggest new causes for the phenomena.

The development of scientific knowledge tells us that without assuming the presence of causal relationships we cannot deduce anything or explain any of the phenomena of nature. Without causality the universe becomes chaotic, where nothing is predictable. However, the most important questions in this topic are: do causes produce their effects directly? Is cause just a representation of the reason for what happens without being effective itself, or is cause actually the agent that is producing the resulting effect?

A cause might be attributed to an innate property (nature) of an object. The question here is: does that property constitute a sufficient cause for the occurrence of the effect? This question has been asked by the Mutakallimūn, for example by al-Bāqillānī in his book, *Tamhīd al-awā'il*, which I will come back to later. If this is the case, as al-Bāqillānī says, then, "the mere availability of irrigation should achieve the growth of plants at all times endlessly, which is not the case."[289] However, one could say that this is not the case because it is an "effect and response" relationship that is needed to expose a cause. Thus, we have here two parts at play: the related, innate property, and the object it is acting upon. Heat, for example, is an innate property, but it does not burn

289 Al-Bāqillānī, *Tamhīd*, 61.

iron as it does paper or cotton. The burning of cotton needs heat and dry cotton. Once the heat is sufficient to set fire to it, the cotton will be ignited and will start to burn by oxidising the carbon molecules in its cellulose fibers. If the cotton is replaced by solid wood, a feeble fire may not be able to ignite it.

We can analyse the above example further. We have two levels of analysis: microscopic and macroscopic. On the microscopic level, we can consider the interaction between the heat source and the cotton molecules, where the heat, in the form of photons or electromagnetic radiation, is absorbed by the carbon atoms (C) in the presence of oxygen molecules (O_2), which is available in the air. This interaction causes electrons in the outer shell of the carbon atom to combine with electrons in the outer shells of two oxygen atoms, which produces one carbon dioxide (CO_2) molecule, and generates, at the same time, an excess amount of heat, light, and some ash; some of the constituents of the cotton. The burning soon becomes spontaneously sustained by a self-generated heat—the basic condition required to burn the rest of the cotton—and a sufficient amount of heat required to excite the carbon atoms, thereby allowing them to combine with the oxygen atoms from the air to form CO_2. This self-sustained reaction continues as long as the conditions do not change. For example, air may become scarce or the cotton may get wet and become difficult to ignite, and so on. In any case, there is no reason why the fire will extinguish unless there is a reason that forces it to do so.

The question here is, what causes the cotton to burn? Is it the heat of an external source which ignites it, or is it that the cotton has a low ignition temperature? Is it because the burning of cotton is an exothermic reaction, which produces enough heat to cause a self-sustained ignition? Is it the availability of oxygen in air? Or is it that oxygen can combine with carbon to form a CO_2 molecule? We can also ask whether it was caused by the atomic structure of either the carbon or the oxygen, or of both. In fact, we can ask an endless number of questions concerning the actual cause for the fire to burn cotton. And it will turn out that many factors play a part in causing this process on the microscopic level.

On the macroscopic level, it would seem that fire plus air plus cotton are the only essential factors for the burning to take place. The fire might be considered the most essential cause for burning, since it is the source of the heat, but nonetheless there are several factors at play in the process of burning cotton, each of which is as essential as the fire itself. This example explains how many factors contribute to the cause of any process in nature. It would be difficult to identify one, and only one, such factor as the cause.

Philosophers admit that it is practically impossible to prove causality, and this is a part of their academic teachings. It is also the reason why it was thought that the occurrence of any event is always associated with apparent conditions. The full conditions cannot all be known while our knowledge about the world is incomplete. From this point of view, it is impossible to say for certain that the cause of any event is fully known, unless all the related parameters are known.

The above simple analysis and discussion is the kind of factual presentation that has to be complemented with a philosophical analysis of the problem. Firstly, we notice that chronological order is a fundamental aspect of causality. Effect occurs chronologically after cause, not the other way around. This may suggest that the effect is generated by the cause, but this is no proof. It could be that such a chronological order presents some conjugation between what has been considered as a cause and the events that follow it.

An example is to say that climatic seasons are caused by the changes occurring in the celestial map of fixed stars. The ancient Arabs would assign climatic events, such as rain, cold spells, or wind, to the appearance or disappearance of constellations from the night sky. Through observing the cyclical commencement of climatic seasons over a long period of time, the Arabs were able to identify a relationship between the appearance of a certain constellation and these events. Since their knowledge of the sky was dominated by astrological beliefs, they concluded that the appearance of constellations must be the cause for the climatic seasons. However, we know now that this was completely false. There is no causal relationship between climatic seasons and the sky map; such a coincidence is a kind of conjugation that occurs because of the motion of the Earth around the Sun. As the Earth moves around the Sun, the view of the sky changes so that different constellations appear in the night sky with different climatic seasons. In fact, this very example is indirect evidence indicating that the Earth makes a full rotation around the Sun in one, full tropical year. Incidentally, one might be surprised to know that the Prophet Muhammad (peace be upon him) warned his followers and prevented them from attributing the cause of such climatic events, rain specifically, to the appearance of certain constellations in the sky, and asked them to attribute it to Allah and take it to be some of His blessings.[290]

One might say that the beliefs of those Arabs were flawed in regard to their basic principles, since it was not possible to prove them scientifically. Although, it might be pertinent to ask, what kind of proof can one provide for it to be an effect other than the verification of the event as it repeats itself regularly over the ages? Well, maybe one needs to prove theoretically that there is a relationship between the configuration of the stars in the sky and the events

290 See the original hadīth in the collections of al-Bukhārī (*Saḥīḥ al-Bukhārī* 4147) and Muslim (*Saḥīḥ Muslim* 125)

on Earth. However, was this not the case with the full astrological organ that dominated scientific thinking at that time? In fact, astrology was part of the teaching curricula in astronomy, while the astrological effect of celestial bodies, including the constellations, was a fundamental part of Greek philosophical doctrines. Was it not the case that Ptolemy, the great astronomer of the Greeks, wrote one of his most important works on astrology? Indeed, as suggested by James Sanders, many beliefs, even those which are scientific, are contextual by nature.[291]

CAUSE IN THE QUR'AN

The Qur'an is the principal source of information for Muslims, and in the early days of Islam the followers of the Prophet (the *Saḥāba*, or "Companions") understood the Qur'an literally. Accordingly, they adopted this understanding, which was carried over to the second and third generations of Muslims. However, it was about a century later that Muslim theologians and thinkers started discussing commentaries of the Qur'an and introduced their interpretations of some of the verses and wordings to conform with rational understanding. Basically, the Qur'an attributes all the events and processes happening in the world, such as rain, wind, disasters, plant growth, the flight of birds, and all acts in the world, to Allah. Here are a few related verses from Surah Al-Anam:

> *It is He Who sends down water (rain) from the sky, and with it We bring forth vegetation of all kinds, and out of it We bring forth green stalks, from which We bring forth thick clustered grain. And out of the date-palm and its spathe come forth clusters of dates hanging low and near, and gardens of grapes, olives and pomegranates, each similar (in kind) yet different (in variety and taste). Look at their fruits when they begin to bear, and the ripeness thereof. Verily! In these things there are signs for people who believe. (6:99)*

There are two points addressed in this verse: the clear affiliation of events with the prime cause, Allah, and the request for humans to pay attention to these signs of Divine action. In Surah Ar-Rad, we read:

> *Allah is He Who raised the heavens without any pillars that ye can see; is firmly established on the throne (of authority); He has subjected the Sun and the Moon each one runs (its course) for a term appointed. (13:2)*

Also, in the same context of natural events we read in Surah Al-Furqan:

> *Have you not seen how your Lord spread the shadow? If He willed, He could have made it still then We have made the Sun its guide. (25:45)*

291 James A. Sanders, *Canon and Community: A Guide to Canonical Criticism* (Philadelphia: Fortress Press, 1984).

One might take this verse as evidence that Allah could rule the universe miraculously in accordance with His wishes, as it would be impossible to make the shadow still unless the Sun is at a fixed point in the sky. Such an act would need a miracle as it seems to go against the natural course of celestial objects. However, this is not the case at all. In fact, the shadow can be brought to a standstill without ordering the Earth to be stationary. To obtain such a state it would be sufficient to prolong the period of Earth's rotation around its axis (one day) to one full year, or to reduce the period of its rotation around the Sun to only twenty-four hours. Perhaps the first option would be better—for a solar day to become a solar year—since the second would mean breaking the laws of mechanics. In this situation, Earth could be said to be in a locked orbit around the Sun. This is an expected situation for some planets that are now being discovered in other solar systems. In a locked orbit, half the globe would always be facing the Sun and the other half facing away from it. The Sun would appear at a fixed point in the sky and, consequently, a shadow over Earth would be constant.

This reminds me of another two verses in Surah Al-Qasas of the Qur'an which are related to this hypothetical situation:

> *Say: Tell me! If Allah made night continuous for you till the Day of Resurrection, which God other than Allah could bring you light? Will you not then hear? (28:71)*

> *Say: Tell me! If Allah made day continuous for you till the Day of Resurrection, which God other than Allah could bring you night wherein you rest? Will you not then see? (28:72)*

In these two verses, Allah is telling us that He could have made the length of the night and the day so prolonged as to continue till the end of life on Earth; which is the Day of Resurrection. But how can this be done without a miracle? In fact, this situation is quite possible and could happen without dispensing a miracle. If Earth were in a locked orbit, then half the globe would be perpetually illuminated by the Sun and the second half would be in darkness. Note that Allah does not just mention the day but the night too and, since both are complementary, half the globe would be in daytime and the other half in night-time. So, as we see from the above two verses, Allah need not do miracles to rule the universe, and the examples given in the Qur'an are for contemplation and to provoke thinking about the phenomena in the world.

In Surah Ar-Rum, we can clearly see the attribution of life and death to Allah; this is one example of the domination of Allah over everything that goes beyond the power and abilities of the human:

> *Allah is He Who created you, then provided food for you, then will cause you to die, then (again) He will give you life (on the Day of Resurrection). (30:40)*

We also read in Surah Luqman:

> *See you not that Allah merges the night into the day, and merges the day into the night, and has subjected the Sun and the Moon, each running its course for a term appointed; and that Allah is All-aware of what you do. (31:29)*

In this verse, once again, the Qur'an stresses the role of Allah on a cosmological scale, but here we have to pay attention to the statement, "*and has subjected the Sun and the Moon, each running its course for a term appointed*", from where one can take the meaning that there is a course that these objects follow for a certain term. Whether this course, or track, is autonomous, or whether it is divinely maintained, is not clear from this following verse in Surah Az-Zumar:

> *See you not, that Allah sends down water (rain) from the sky, and causes it to penetrate the Earth, (and then makes it to spring up) as water-springs and afterward thereby produces crops of different colours, and afterward they wither and you see them turn yellow, then He makes them dry to become chaff. Verily, in this, is a Reminder for men of understanding. (39:21)*

In other verses, the Qur'an points to angels as being Allah's messengers. They are sometimes directed to undertake certain duties and to execute the orders of Allah. For example, in Surah As-Sajdah, angels execute the order of Allah on death:

> *The angel of death, who is set over you, will take your souls, then you shall be brought to your Lord. (32:11)*

In addition, in Surah Al-Imran, it is mentioned that the angels called upon Maryam with a message from Allah concerning her privileged status:

> *And when the angels said: O Maryam, surely Allah has chosen thee and purified thee and chosen thee above the women of the world. (3:42)*

We also read in Surah Al-Anfal, that angels were sent to fight side-by-side with the believers in support of their fight against the non-believers at the Battle of Badr:

> *(Remember) when you sought help of your Lord and He answered you (saying), "I will help you with a thousand of the angels each behind the other (following one another) in succession." (8:9)*

It is also understood, from the Qur'an, that angels are always being employed to do their duties in Heaven (21:103) and in Hell (74:31). It is known that angels were sent to Lūṭ's city to destroy it, and on their way, they passed by Ibrahim (Abraham) and told him of the good news that he would have a son

(11:69–83). Therefore, we can confidently conclude that the angels are servers who carry out the orders of Allah; accordingly, they are a means through which Allah executes His orders. The nature of the angels and how they work remain mysteries, but certainly they are something to be resolved by theologians, one day.

My point here is to conclude that the Qur'an does not object to the existence of secondary causes, only that such secondary causes are to be understood as the means of execution, no more; meaning that they cannot take decisions of their own or object to a divine order. Here, it might be important to point to a mistake made by Harry Wolfson when he states that, "in one instance an angel is said to have disobeyed God's command (2:28–32; 38:71–74)."[292] In fact, in Islamic tradition, it was not an angel who disobeyed the divine order, but the Devil, who was a *jinn* (see Qur'an 18:50). Angels are obedient creatures that can, by no means, disobey Allah, since they were originally created to obey, as we see in Surah At-Tahrim:

> *Over which are (appointed) angels stern (and) severe, who disobey not, (from executing) the Commands they receive from Allah, but do that which they are commanded. (66:6)*

Most Muslim theologians and thinkers followed the basic presentation of the Qur'an concerning the understanding of causality and asserted the belief that Allah is the only cause in the world and everything that happens in it. It is notable that scholars from both schools of *kalām*: the Ashʿaris and many of the Muʿtazilis (excepting two characters), approved of this understanding and took it as part of the basic beliefs of Islam. Wolfson[293] has discussed this point in great detail and resolved some questions, which are of historical interest. The reason for such an insistence on this concept is the fact that the beliefs of Islam require one not to assign any partner to Allah in His creation and ruling of the world. This is the absolute essence of *tawhīd*. If any kind of partner is assigned, whether implicit or explicit, then the person believing this is no longer a Muslim. This is why Muslims could not tolerate the assumption that nature can act on its own, or that causes can affect events, intrinsically, by their very nature. This leads to *shirk* (idolatry or polytheism), which would disqualify one from being a Muslim. This may explain why a great thinker like al-Ghazālī could not accept causal nature or deterministic causality, but, instead, insisted on the absolute freedom of the Divine to affect anything imaginable in the world, including turning a book into a "beardless slave boy."[294]

---

292 Wolfson, *Kalām*, 519.

293 Ibid. 559.

294 Al-Ghazālī, *Incoherence*, 170.

When taken as part of public belief, the denial of deterministic causality became a factor that hindered scientific thinking. Spread through the masses, such a belief effectively denied the existence of a law that controls the workings of all the world's phenomena. This aborted rational thinking encouraged laziness and a belief in the Utopian state of the Islamic mind. Many of the verses in the Qur'an that encourage scientific thinking were ignored, and it was thought that the universe ran miraculously. This was a very dangerous transformation in the Islamic media, taking intelligent theologians away from subjects that they should have considered. What makes the situation worse are the restrictions that have been imposed on theological studies and religious innovations (*ijtihād*) since the 15th century. No innovative studies were allowed, except those that were in line with the traditional belief system. These restrictions, along with other factors, arrested the Islamic mind and wasted Islamic genius. Indeed, this was one of the fundamental reasons for a decline in Islamic civilisation. This period of decline in the Islamic world has extended from the 15th century onward.

CAUSALITY ACCORDING TO TRADITIONAL KALĀM

The early Mutakallimūn were seriously engaged with this problem, which was one of the main issues in their endeavour to explain and understand the world according to Islamic beliefs. In a joint article, some colleagues and I examined the concept of causality, both in *kalām* and modern physics, and showed that there are some aspects where the *kalām* view shared the modern scientific view on this topic.[295] Here, I will try to clear much of the mist concerning the concept of causality in traditional *kalām*, in order to achieve justice for all concerned parties.

The general consensus is that the Muʿtazilis believed that the world is causally deterministic, whilst the Ashʿaris hold that it is occasionalistic. We will see that this is, in fact, not true and that both parties actually subscribe to the same basic view with respect to natural causality, as well as their understanding of the actual causes and causal relationships between events in the world. We will see that exponents of both sides agree on the same designations concerning the essence of a cause and attribution of causes to Allah, who is affecting them. The only topic where the Muʿtazilis and the Ashʿaris disagree is the question of human causality.

The Mutakallimūn adopted the stipulations of the Qur'an directly with regards to their belief in divine omnipotence and omniscience. They accepted that Allah is the only actual cause effecting changes in the world; He is the

295 Mohammed Basil Altaie, A. Malkawie, and M. Sabbarini, "Causality in Islamic *Kalām* and in Modern Physics", *Jordan Journal of Islamic Studies,* Vol. 8, 2A (2012). Arabic

Creator and the Generator of all events. They had differing opinions on the efficacy of the intrinsic nature of things and its role in effecting a cause. In principle, the Mutakallimūn rejected the idea that the innate properties of an object constitute a nature that can act autonomously to effect change. Instead, they attributed such an act to Allah and, to be more precise, some of them assumed that Allah has, by His own obligation, accepted the role played out through the nature of things in effecting the cause. Most of the Mutakallimūn who adopted such views belonged to the Baghdadi group, i.e. followers of Abū al-Qāsim al-Kaʿbī. Generally, all these views can be termed "occasionalist." The philosophical basis for the views of the Ashʿaris was drawn in accordance with their proposal of "re-creation." This is one of the basic principles of *Daqīq al-kalām*, which were presented in Part I of this book. They drew this principle from their understanding of the most fundamental attribute of Allah, i.e. that of being the Creator. Likewise, as Allah is active in the universe at all times and He is described in the Qur'an as being the dominant actor who sustains the universe, the Mutakallimūn interpreted this attribute of Allah, the Sustainer, through the principle of re-creation. Accordingly, all the implications of this assumption were used to explain the physical world.

The Ashʿaris and almost all of the Muʿtazilis agreed that Allah is the prime cause for whatever happens in the physical world. Both schools deny that the innate properties (*ṭabʿ*) of things are able to act autonomously to cause a change in the world. For this reason, they denied the effective nature of things. However, they had different views about the details of how the Divine effects cause and, further, disagreed over the question of human free will and responsibility for their own actions.

VIEWS OF THE MUʿTAZILIS

The Muʿtazilis formulated a sophisticated theory of causality that preserved the basic Islamic belief that Allah is omnipotent and can intervene to enforce His will. There are quite a number of works that are very useful in providing a picture of the Muʿtazilis' views on causality and causal relationships, for example, the book of Qāḍī ʿAbdul Jabbār (d. 1024), *al-Muḥīṭ bi al-taklīf*, and his celebrated encyclopeadia, *al-Mughnī*. In addition, we find al-Khayyāt's *al-Intisār* quite a useful source, in which we find many quotes from al-Naẓẓām. The work of Abū Rashīd al-Naysābūrī, *Masāʾil al-khilāf*, is a good source for understanding the details of the differences among the Muʿtazilis themselves, whereas the book of Ibn Matawayh, *al-Tadhkira*, is more concerned with clarifying the concept of adherence (*iʿtimād)*, and its implications with respect to causality when discussing the finer arguments in *kalām*.

Analysing the legacy of the Muʿtazilis on causality, I gather that they identified four types of secondary causal relationships: adherence (*iʿtimād*), conjunction (*iqtirān*), generation (*tawlīd*), and custom (*ʿādah*). These four terms were used to describe different causal relationships in the world. I have shown, in Chapter Six of this book, that Qāḍī ʿAbdul Jabbār has rejected nature and natural action, and attributed action directly and indirectly to Allah. He accepted that an object may exhibit adherence (*iʿtimād*) for certain actions, say, due to its position, in which case it could respond to external effects. As such, this adherence may then play the role of an intermediate cause. ʿAbdul Jabbār said:

> And when they [the philosophers] mean by 'Nature' that which happens from the burning fire, it is what we consider to happen out of the adherences (*iʿtimādāt*), which generate dissociations. This is as if they call *Nature* what we call *iʿtimād*. This is similar to what happens to weight upon immersing a heavy [body] in water, generating a fall, and similar things. We can attribute all these [events] to an actor who is free and able to prevent such a generation and response.[296]

Here, ʿAbdul Jabbār differentiates between two types of causal relationships: the first is when Allah is directly interfering as the cause of the event (this is identified as *ʿādah*), and the other is caused by the indirect intervention of Allah (which is identified as both *iʿtimād* and *tawlīd*). ʿAbdul Jabbār goes further in his explanation to show that the effect of becoming drunk from drinking wine and spirits can be interpreted as a custom (*ʿādah*), which is a concept that was also used by the Ashʿaris and which will be explained later in this chapter. ʿAbdul Jabbar says, "if they would include in this what happens of becoming drunk when drinking [wine], then this follows the custom (*ʿādah*) and it is God which makes it without having a necessary cause (*amr yūjib*); this is because if drinking would cause drunkness, then it would happen even with drinking water... and all this is flawed. This demonstrates that what occurs does not belong to causes and natures but follows the custom."[297] As we will see shortly, in using the same examples, what we read above is a repetition of the views of al-Bāqillānī.

As far as what may be understood as secondary causes, such as generation (*tawlīd*), ʿAbdul Jabbar points out that such generated effects are produced by the acts of man and not by the acts of the inanimate, which are normally associated with the adherences (*iʿtimādāt*). In this respect, ʿAbdul Jabbar objects to what al-Jāḥiz says regarding those events generated by the will of a person,

296 ʿAbdul Jabbār, *Al-Muḥīṭ bil taklīf*, 101.
297 Ibid.

where he claims that such a will happens by nature. 'Abdul Jabbar refutes this, saying that there is no difference in occurrence by nature, whether it is accounted for on the part of the will or by the event itself, so rejecting what is being claimed to happen by its own nature.[298]

Some prominent Muʿtazili scholars claimed to have affirmed causality in a way that may appear to agree with the concepts of the philosophers. Wolfson has spent a great deal of effort proving that Muʿammar and al-Naẓẓām affirmed causality,[299] however, I do not think that this claim is accurate enough. Al-Naẓẓām, for example, although he agreed that objects have some kind of nature that qualifies them to interact with other objects in a systematic (lawful) way, is found to have subscribed to the idea that it is in the power of Allah to prevent a causally expected event. Al-Khayyāṭ quotes al-Naẓẓām, saying that, "if it was the quality of water to flow and if it was the property of a heavy stone to fall, then Allah is able to prevent it from doing so."[300] This intimates that al-Naẓẓām, although admitting that effects are carried through the innate qualities associated with things, acknowledges that it is in the power of Allah to prevent the stone from falling; in this he is bluntly denying causal determinism. On the other hand, Muʿammar al-Sullamī, who accepted the atomic theory of *kalām*, suggests that nature follows causally deterministic laws without any intervention from Allah, except for the first creation. He believed that objects have intrinsic natures which have the power to generate accidents (*aʿrāḍ*) without divine intervention.[301] This opinion of Muʿammar was harshly criticised by al-Shahrustānī,[302] ʿAbdul Jabbār,[303] and Ibn Ḥazm.[304]

Concerning adherence, ʿAbdul Jabbār refutes Muʿammar and al-Naẓẓām, saying that, "those accidents which persist in the inanimate or else exist by their nature, such as the movements of the stone and similar accidents, where they say that none of these are related to God except for his allocating it, but it is what is necessitated by its own nature." Additionally, he says that they (Muʿammar and al-Naẓẓām) have, "discarded the generated effects from being the acts of man and attributed them to God by the necessity of man's creation or attributed it to be the act of the self by its own nature."[305] This is a very important discussion that ʿAbdul Jabbar is presenting here, as it reflects his

298 Ibid. 305.

299 Wolfson, *Kalām*, 559–78.

300 Al-Khayyāṭ, *Kitāb al Intiṣār wa al Rad ala Ibn al-Rawandī*. 1988: edited by M. Hijāzī, al Thaqāfah al Dīniyah Bookshop, Cairo, 48.

301 Pines, *Studies in Islamic Atomism*, 31.

302 Al-Shahrastānī, *milal*, 65–8

303 ʿAbdul Jabbār, *Al-Muḥīṭ bil taklīf*, 385.

304 Ibn Ḥazm, *Fiṣal*, 58–9.

305 Ibid. 386.

agreement with the Ash'aris on this matter. To counter the claims of Mu'ammar and al-Naẓẓām, he says, "according to this, God would not be considered as acting on any of the accidents originally… and this view of al-Naẓẓām implies that God would be committing an injustice…."[306] In fact, this shows clearly how the prominent Mu'tazili, 'Abdul Jabbar, and his school regarded the question of natural causality.

VIEWS OF THE ASH'ARIS

From the Ash'ari perspective, I present here the views of al-Bāqillānī, the famous Ash'ari theologian who discussed the concept of nature and causality in his book, *Tamhīd al-awā'il wa talkhīṣ al-dalā'il.* In his discussion, al-Bāqillānī tried to disprove three of the proposals which were known in philosophy:

1. that the world was produced out of a certain nature that made it necessary to exist once it was available.
2. that the world is made of the four basic elements: fire, air, earth, and water.
3. that celestial objects have no effect on the Earth, which means that astrology is null and void.

The essence of al-Bāqillānī's argument is based on the requirement that an action by nature would be possible only if nature has a will and the ability to choose, for otherwise no action is expected from a blind inanimate nature. This argument found favour with several Ash'ari theologians and was also adopted by some of the Mu'tazilis, such as 'Abdul Jabbār, as shown above and in Chapter Four.

THE ASH'ARI THEORY OF CUSTOM (*'ĀDAH*)

The Ash'aris interpreted causal relations and causal determinism in a flexible manner, saying that the laws of nature are the sorts of relations that we are accustomed to seeing manifested in natural phenomena. This means that the apparent causal relationship found between causes and their effects are not necessary and it is simply that the world is designed to behave that way. The Ash'aris saw no meaning in causal relationships, as long as an active role was not assigned to the cause itself, and, since Allah is the direct cause for what happens, therefore, causal relationships are no more than a formal expression of the actions behind the phenomena. They accepted that Allah assigned this kind of formality to rule the universe. Accordingly, miracles were thought to be an infringement of these customs. There seems to be no known mechanism for the custom to occur, or for how such a custom was brought into existence.

306 *Loc.cit.*

This is one of the main arguments that makes one wonder about the impact of such an apologetic approach to understanding the world. I would rather say that the concept of custom, although used by the Ashʿaris and Muʿtazilis, is not a profound concept that can explain anything at all. It would have been better to use the expression "Sunna of Allah" (meaning, the enactment of Allah) instead of *ʿādah*, since it is implied that the laws by which the universe is ruled are the laws of Allah.

AL-GHAZĀLĪ ON CAUSALITY

In the seventeenth discussion of his book, *The Incoherence of the Philosophers*, al-Ghazālī denied that causes are the direct reasons for events occurring in nature or that causal connections are necessary for events to happen:

> The connection between what is habitually believed to be a cause and what is habitually believed to be an effect is not necessary, according to us. But [with] any two things, where "this" is not "that" and "that" is not "this," and where neither the affirmation of the one entails the affirmation of the other nor the negation of the one entails negation of the other, it is not a necessity of the existence of the one that the other should exist, and it is not a necessity of the nonexistence of the one that the other should not exist—for example, the quenching of thirst and drinking, satiety and eating, burning and contact with fire, light and the appearance of the Sun, death and decapitation, healing and the drinking of medicine, the purging of the bowels and the using of a purgative, and so on to [include] all [that is] observable among connected things in medicine, astronomy, arts, and crafts.[307]

He argued that causal dependence is far from being trivial and that it would need rigorous proof that causes are the necessary and sufficient reasons behind their effects. Al-Ghazālī discussed the example of the burning of cotton and argued that the innate properties of objects cannot be the cause of events, since such properties (such as fire) are inanimate and have no "action", "As for fire, which is inanimate, it has no action. For what proof is there that it is the agent? They have no proof other than observing the occurrence of the burning at the [juncture of] contact with the fire. Observation, however, [only] shows the occurrence [of burning] at [the time of the contact with the fire], but does not show the occurrence [of burning] by [the fire] and that there is no other cause for it."[308] Clearly, al-Ghazālī did not consider observation as decisive evidence, since such evidence is no proof of an actual effective cause but is merely evidence of the conjunction between what is thought to be a cause and what

307 Al-Ghazālī, *Incoherence,* 166.
308 Ibid. 167.

is called, an effect. In a similar mode to the arguments given above concerning the necessary and sufficient conditions for the cause, al-Ghazālī insisted that we should claim the sufficiency of evidence as:

> Whence can the opponent safeguard himself against there being, among the principles of existence, grounds and causes from which these [observable] events emanate when a contact between them takes place [admitting] that [these principles], however, are permanent and never ceasing to exist; that they are not moving bodies that would set; that were they either to cease to exist or to set, we would apprehend the dissociation [between the temporal events] and would understand that there is a cause beyond what we observe? This [conclusion] is inescapable in accordance with the reasoning based on [the philosophers' own] principle.[309]

In addition, al-Ghazālī brought into the analysis the question of the interaction between the causal object and the affected object. Here, he discloses the fact that, for some unknown reason, the cause might not act to produce the expected result. For example, "a person who covers himself with talc and sits in a fiery furnace is not affected by it."[310] Through this example, al-Ghazālī tried to rationalise the possibility of miracles, saying that, even if we admit, "Fire is created in such a way that, if two similar pieces of cotton come into contact with it, it would burn both, making no distinction between them if they are similar in all respects. With all this, however, we allow as possible that a prophet may be cast in the fire without being burned, either by changing the quality of the fire or by changing the quality of the prophet."[311]

However, in this respect, al-Ghazālī's argument seems weak. He could have used the stronger argument of the Mutakallimūn, which is the argument of re-creation by which it would be legitimate to see a possibility, at least on the theoretical level, of miracles happening. Instead, he resorts to unknown conditions for the occurrence of an effect:

> If then, the principles of dispositions are beyond enumeration, the depth of their nature beyond our ken, there being no way for us to ascertain them, how can we know that it is impossible for a disposition to occur in some bodies, which allows their transformation in a phase of development in the shortest time so that they become prepared for receiving a form they were never prepared for receiving previously, and that this should not come about as a miracle? The denial of this is only due to our lack of capacity to understand, [our lack of] familiarity with exalted beings, and our unawareness of the secrets of God, praised be He, in creation and

309 Ibid. 168.

310 Ibid.

311 Ibid.

> nature. Whoever studies inductively the wonders of the sciences will not deem remote from the power of God, in any manner whatsoever, what has been related of the miracles of the prophets.[312]

Al-Ghazālī concurred with the Ashʿaris' view in denying causal determinism and attributed the observed causal relations to a custom (*ʿādah*). I consider the above argument on causality to be incomplete. However, the most prominent points in al-Ghazālī's argument are that innate properties are inert and have no actual effect by themselves, and that causes are only known whenever all necessary and sufficient conditions in the relationship between the cause and the effect are known. The most important element in his argument, which I call the "minimal principle on causality", is that causal determinism is denied. This basically stems from the Islamic refutation of determinism in nature and is one of the basic principles of Islamic *kalām*.

It is sometimes argued that al-Ghazālī accepted the existence of intrinsic nature when he described the innate properties of things. This is true, and indeed one can find him defining lightness and weight, saying, "lightness is a natural force by which the body moves away from its position by nature, and weight is a natural force by which the body moves to its position by nature."[313] He also defined moistness as, "a responsive quality by which the body accepts bonding and forming easily without preserving it, but retaining its shape and status as per the motion of its nature."[314] In these definitions, I assume that al-Ghazālī was following Aristotle. However, elsewhere he states, "The basic point regarding all of them is for you to know that nature is totally subject to God, Most High: it does not act by itself but is used as an instrument by its Creator. The sun, moon, stars, and elements are subject to God's command: none of them effects any act by, and of, itself."[315] Clearly then, there is no question about the intentions of al-Ghazālī on using the word "nature" or his understanding of its role in serving the cause.

To summarise the position of the Mutakallimūn on causality and causal relations, I can identify the following:

1. They denied the action of any sort by intrinsic nature of things (*ṭabʿ*).
2. They denied causal determinism, assuming that the regularity of causal relationship is only a sort of custom (*ʿādah*).
3. They acknowledged the law and order observed in the behaviour of the world, but allowed for miracles to occur through the infringement of custom.

---

312 Ibid. 174.

313 Al- Ghazālī, *Mi'yar*, 174.

314 Ibid.

315 Al-Ghazālī, *Deliverance from Error*, 10.

Besides the above short explanation, I should remark here that an excellent analysis about al-Ghazālī and his position on causality can be found in Frank Griffel's book, *Al-Ghazālī's Philosophical Theology.*[316]

IBN RUSHD'S CRITICISM OF THE ASH'ARIS' CUSTOM

Ibn Rushd (Averroes) criticised al-Ghazālī's *Tahāfut al-falāsifah* (*The Incoherence of the Philosophers*) in a work entitled, *Tahāfut al-tahāfut* (*The Incoherence of the Incoherence*). In this text, Ibn Rushd remarked that, "to deny the existence of efficient causes, which are observable in sensible things, is sophistry."[317] He added that the causal properties of an entity are an essential aspect of our understanding of that entity. That is, if we try to strip things of their causal properties to reveal what "real" substance there may remain, we end up with nothing at all. This is because the ways in which we label objects are directly influenced by our grasp of their characteristics vis-à-vis other objects. Anyone who casts doubt on causation damages the quest for knowledge, as causation is intimately connected with our knowledge of the world. To quote Ibn Rushd:

> Logic implies the existence of causes and effects, and knowledge of these effects can only be rendered perfect through knowledge of their causes. Denial of causation implies the denial of knowledge, and denial of knowledge implies that nothing in this world can really be known, and that what is supposed to be known is nothing but opinion, that neither proof nor definition exists, and that the essential attributes, which compose definitions, are void.[318]

Ibn Rushd further argued that even though the ultimate cause of every phenomenon is God, He has also established a secondary cause for every phenomenon. While God is capable of bringing about satiety without eating, quenching of thirst without drinking, and burning without contact with fire, He does not normally do so, and when He does so, it becomes a miracle, which is among the, "Divine acts and beyond the reach of human intellect."[319]

In several pages, Harry Wolfson presented what he called, the "five arguments of Averroes",[320] against the denial by the Mutakallimūn of nature and causality. Although Wolfson was very keen on referring every sentence made by Ibn Rushd to Aristotle, I could not find a profound argument in his criticism for the denial of causality that would answer the question of whether "nature" and "natural properties" have qualities, power, or choice to effect a

316 Frank Griffel, *Al-Ghazālī's Philosophical Theology* (Oxford and New York: Oxford University Press, 2009).

317 Averroes, *Tahāfut al-Tahāfut*, 587.

318 Ibid, 584.

319 Ibid, 587.

320 Wolfson, *Kalām*, 551–8.

change, autonomously. This is the basic question in respect of causality. To say that God works through secondary causes is not an issue for denying that the innate properties of a thing can affect incidences in the world which result in world-changing events. On the other hand, a close look at the arguments of the Mutakallimūn shows that they accepted the existence of law and order in the world. The real issue is whether actions in the world can be attributed to those entities (properties, nature) rather than Allah.

LAPLACE'S DETERMINISM

Pierre Simon Laplace (1749–1827) lived in an era of enormous change and innovation. He actively contributed to the development of Newtonian mechanics and paved the way for further works that made the Newtonian picture of the world flourish. Laplace realised that this picture offered a deterministic universe by which we could predict every future, once the initial conditions of the system were known:

> "We ought to regard the present state of the universe in terms of the effect of its antecedent state and as the cause of the state that is to follow. An intelligence knowing *all* the forces acting in nature at a given instant, as well as the momentary positions of *all* things in the universe, would be able to comprehend, in one single formula, the motions of the largest bodies as well as the lightest atoms in the world, provided that its intellect were sufficiently powerful to subject *all* data to analysis. To it, nothing would be uncertain, the future as well as the past would be present to its eyes. The perfection that the human mind has been able to give to astronomy affords but a feeble outline of such intelligence."[321]

This is, perhaps, the best description in physics for the deterministic structure of classical mechanics. As one can easily see, this determinism requires knowing all the forces acting in nature and the momentary positions of all things in the universe; something that cannot be guaranteed. However, taking the possibility of such knowledge as an assumption, one can see that Laplace, in fact, referred to a super-being that is, "sufficiently powerful to subject *all* data to analysis; to it nothing would be uncertain, the future as well as the past would be present to its eyes." This affirmation underlies a kind of metaphysical speculation over the presence of an omniscient entity. This does not mean, of course, that Laplace either believed in or denied the presence of such an entity, but we can be sure that he denied the intervention of any supernatural power in the ruling of the universe through his answer to Napoleon's remark about the Creator, by saying, "Je n'avais pas besoin de cette hypothèse-là" (I had no need of that hypothesis).[322]

---

321 Laplace, *A Philosophical Essay on Probabilities*.

322 Walter W. Rouse Ball, "Pierre Simon Laplace (1749–1827)", in *A Short Account of the*

Generally, all the laws of physics are deterministic, since they are represented in the form of differential equations. However, this does not mean that these laws will describe, with absolute accuracy, the laws of nature. As I have shown in Chapter Five, the laws of nature govern natural phenomena, whereas the laws of physics are the invention of our minds, formulated within the capacity of our logic. Historically, there was a lot of confusion regarding this concept, the result being that the deterministic laws of physics were taken to imply a deterministic nature. This is not true; we have the Schrödinger equation, for example, which is deterministic, and describes the most basic equation in quantum mechanics, but we now know that quantum nature is probabilistic.

CAUSALITY IN THE THEORY OF RELATIVITY

In 1905, Albert Einstein proposed the Theory of Special Relativity as a resolution for some of the fundamental problems in classical physics. He proposed that space and time form an integrated continuum, and that the velocity of light in a vacuum is a universal constant, independent of the state of motion of either the observer or the source. With this assumption, Einstein was able to resolve the problem of the propagation of light in a vacuum without the need to make use of the hypothetical medium, known as "ether." Space and time became one integrated entity, now called "spacetime." Consequently, a new concept of simultaneity was developed by which the minimum time needed to propagate an effect through space, from one point to another, is equal to the spatial separation between the two points divided by the velocity of light. No effect can be transmitted in a shorter duration of time because the fastest signal that can carry the effect is the velocity of light in a vacuum, which is about 300,000 km per second. If any effect is thought to be propagated faster than this, it would be considered non-causal. Therefore, a new understanding of physical causality was established by the Theory of Relativity, and further developed by redefining simultaneity and simultaneous events as being limited by the finite velocity of light.

Hermann Minkowski described flat spacetime in terms of causal and non-causal regions (see Figure (3) below). His famous spacetime diagrams considered to be causal only those events that fall within the so-called past and future light cones. Any point (event) outside these light cones is deemed to be non-causal, since it would need a signal faster than light to deliver it. As seen in the diagram, the whole of spacetime is composed of two main regions: two time-like regions, the past and the future, and two space-like regions. In the time-like regions, all events are connected by signals which travel with the

History of Mathematics, 4th ed. (New York: Dover Publications, 2003).

speed of light, or slower. This is why they are identified as being causal, where causes and effects obey a chronological order.

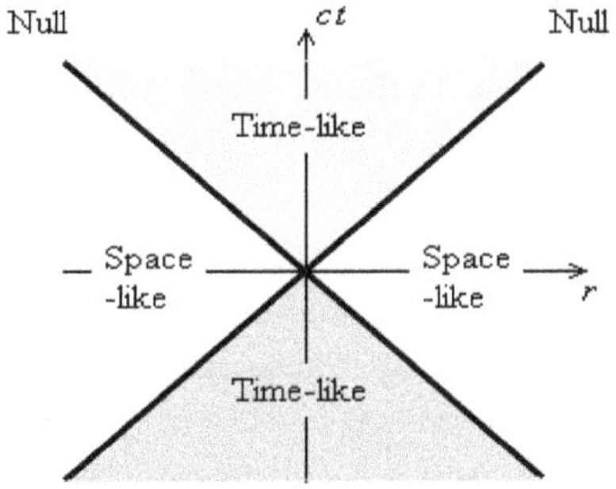

Figure (3) The Light Cone

The space-like regions in the diagram are non-physical, with events needing a signal faster than light to connect them, meaning that they are non-causal, as causes and effects do not necessarily respect chronological order. In fact, it could be imagined that, in a purely spatial world, time does not exist and we can identify the development of things only from their progress through space. Although a difficult concept to imagine, it could perhaps be comprehended by thinking of spatial variations which replace temporal variations. In a space-like world, you are always on the move, transferring from one position to another. With such a continuous change in space, you are proceeding through space and discovering its content.

Relativity theory was used in particle physics to explain some non-causal events. This was possible by assuming the existence of particles moving with velocities larger than the velocity of light in a vacuum. Such a state is possible only if the particles have an imaginary rest mass. These particles are called tachyons. However, the problem with this is that an imaginary mass cannot be physically measured. For this reason, we cannot consider tachyons to be real particles. If, however, we get back to the fact that spacetime is composed of two parts, and assume that these parts are complementary to the extent that the physics of our time-like world might be affected by objects and events taking place in the space-like world, then we might be able to understand many aspects of our world more profoundly and coherently. This includes, for example, the phenomenon of quantum entanglement, which will be explained in the next section, and which involves some sort of "spooky action at a distance", as described by Albert Einstein.[323]

323 This came in a letter from Einstein to Max Born, dated 3 March 1947 (*The Born-Einstein Letters*; Correspondence between Albert Einstein and Max and Hedwig Born from 1916 to

There are many other phenomena that need to be understood in a wider context, in which spacetime is considered with its entire content so that one might be allowed to see all possible effects and entities. For now, we might see only half the facts of our world. The two aspects of spacetime are complementary and what was considered a "spooky action at a distance" might be explained by analysing the space-like regions of spacetime. There are several such enigmas in contemporary physics.

The Theory of General Relativity provided us with a broad arena for considering the integration of space and time on a more profound level. In this theory, spacetime is preserved. Spatial variations are always complemented by temporal variations, and vice versa, so as to keep a specified spacetime interval constant. The physics of compact objects and black holes cannot be understood without the laws of general relativity. The understanding of these objects might be relevant for an appreciation of our future; a huge black hole might very well be the fate of our universe. Nevertheless, there are some mysteries concerning these objects. For example, we are not sure whether a real singularity exists at the centre of black holes or not; it is also not clear what sort of a world is to be found inside the event horizon, nor is it clear whether objects falling into a black hole are lost forever. The most recent declaration made by Stephen Hawking,[324] in which he says that the event horizon might not be a one-way path, is not much help since he is still thinking of physics within our time-like universe. Again, in order to understand the full picture, I think it would be necessary to go beyond the time-like universe. This is inevitable, since the region inside the event horizon of a black hole is space-like. Therefore, we can confidently affirm that the concept of causality in the Theory of Relativity is not completely clear, despite the fact that this theory established a well-defined border between the causal and non-causal regions of spacetime.

CAUSALITY IN QUANTUM MECHANICS

Quantum mechanics is the mechanics of the indeterminate nature. It is the best theory known, to date, that describes matter and energy, and their mutual interactions. The quantum nature of matter and energy has exposed the fact that our knowledge about the world is associated with an inherent uncertainty. This is described by the Heisenberg uncertainty principle, which states that the product of inherent uncertainty in the position of a particle multiplied by

---

1955, Walker, New York, 1971); cited in M.P. Hob-son *et al.*, *Quantum Entanglement and Communication Complexity* (1998), 1–13.

324 Stephen Hawking, "Information Preservation and Weather Forecasting for Black Holes" (lecture, Fuzz or Fire Workshop, The Kalvi Institute for Theoretical Physics, Santa Barbara, CA, August 2013), preprint available at arXive: 1401.5761v1, accessed 22 June 2013.

the inherent uncertainty in its momentum must be greater than half the value of Planck's constant (ħ/2). This means that we cannot determine the position of a particle and its speed with absolute certainty. This refutes Laplace's determinism, which states that the momentary positions of particles and the forces acting in the universe can no longer be determined accurately enough to verify them. Therefore, the universe is no longer deterministic as it would be impossible to designate the position of a particle and its momentum, simultaneously, with infinite accuracy.

This discovery shook the very foundations of classical physics. In order to know the exact trajectory of a comet in the sky, for example, we need to know, with a high degree of accuracy, its position and speed at a given time. Without highly accurate initial data, we are uncertain of the possibilities of its trajectory. The consequences of uncertainty become even worse with nonlinear systems, where inaccuracies in the initial conditions might cause highly unpredictable results.

Quantum mechanics has demonstrated that the occurrence of natural phenomena is probabilistic rather than deterministic. Let me explain the quantum impact on the concept of causality. The availability of physical conditions (the causes), which are required for a physical phenomenon to occur, does not guarantee that the phenomenon will certainly occur. For example, the emission of nuclear radiation (radioactive decay) is a probabilistic event that cannot be guaranteed to occur once the required nucleus appears with the required properties.

The probabilistic behaviour of quantum systems has led to the problem of measurement in quantum mechanics. This problem is reflected in the fact that a measurement outcome is not known for certain unless a measurement is made. This has led to some paradoxes, such as that of Schrödinger's cat[325] and the Einstein—Podolsky—Rosen (EPR) paradox.[326] In the case of the Schrödinger's cat paradox, the principles of quantum mechanics stipulate that the state of a cat inside a closed box, together with a radioactive source and a poisoning device, which can be initiated by the radioactive source, is unknown. In fact, it is a mixed state: the cat might be alive *or* dead, with probabilities dependent on whether the radioactive source emits radiation or not. This sort of indeterminism leaves uncertainty over the occurrence of the phenomenon, despite all elements being present. From this point of view, causal events no longer have a deterministic nature. However, this does not mean that

325 Erwin Schrödinger, "Die Gegenwärtige Situation in der Quantenmechanik [The Present Situation in Quantum Mechanics]", *Naturwissenschaften* 33 (November 1935).

326 Albert Einstein, B. Podolsky, and N. Rosen, "Can Quantum-Mechanical Description of Physical Reality Be Considered Complete?", *Physical Review* 47 (1935): 777–80.

an event in the quantum world would be non-causal, certainly not, but it is the causal determinism that is abolished here and not causality itself. Nevertheless, there remains the question of whether causes themselves are the reasons by which the events might happen or not. As far as I can see, neither science nor philosophy has given a sensible resolution for this problem.

In the EPR paradox, we have another form of causality breakdown; this is spacetime causality. In this phenomenon, two quantum states, separated from one another by a large distance, are entangled in such a way that a change in the physical properties of one would cause an immediate change in the corresponding properties of the other, irrespective of the distance between them. This is a sort of "spooky action-at-a-distance", since we acknowledge as being causal those events for which effects occur after a time duration, that has to be equal or larger than the distance between the cause and its effect divided by the velocity of light. However, in light of the discussion in the previous section, we may speculate that such an event, for which the time duration of transmission is less than the distance divided by the velocity of light, occurs through some sort of a space-like channel. This means that the signal from one point to another is transmitted through a space-like channel.

The indeterminism of nature, as presented by quantum mechanics, is preserved throughout the time evolution of quantum systems, which is covered by the so-called unitary transformations. However, this indeterminism is not an artefact of the theory but is something that is inherent in natural phenomena. Some people do not realise that the laws of quantum mechanics are a form of empirical law rather than being a derivative of a theory. Most of the quantum descriptions of natural phenomena in the microscopic world are the products of observations, whereas theory merely allows for an explanation of the data that an experiment can provide. For example, the observation that particles behave like waves is an experimental fact that has been demonstrated using particle diffraction experiments. The phenomena of electrons tunnelling through potential barriers with energies higher than their own has been observed in semiconductor devices; without these our mobile phones and computers would not work. Even the superposition of states and their non-locality, entanglement, and other bizarre phenomena have all been observed in laboratories.[327] Thus, it would appear to be out of the question that indeterminism could be subjected to invalidation or replacement by a deterministic theory.

327 Gilbert Grynberg, Alain Aspect, and Claude Fabre, *Introduction to Quantum Optics: From the Semi-Classical Approach to Quantized Light* (New York: Cambridge University Press, 2010). In this book, many experiments are described in which the pure quantum aspects of light and particles are exposed.

The hidden variables theory, which was suggested by David Bohm,[328] is not a viable replacement for the indeterministic quantum description. The apparent statistical nature of quantum measurements has led some people to think that one day, a new theory might be discovered by which classical determinism and causality is restored, although such a dream seems to be far-fetched. In John Wheeler's words, "[A]s the statistical character of quantum theory is so closely linked to the inexactness of all perceptions, one might be led to the presumption that behind the perceived statistical world there still hides a "real" world in which causality holds. But such speculations seem to us, to say it explicitly, fruitless and senseless... quantum mechanics establishes the final failure of causality."[329] It has been already proved by John Bell that non-local hidden variable theories are out of the question and have no real validity.[330] On the other hand, no single known prediction can be made to promote hidden variable theories over the standard formulation of quantum mechanics.

Some of the most prominent physicists and philosophers believed in the determinism of the world, including Paul Dirac who, despite being one of the founders of quantum physics, expressed his belief in a deterministic universe:

> It seems clear that the present quantum mechanics is not in its final form. Some further changes will be needed, just about as drastic as the changes which one made in passing from Bohr's orbits to quantum mechanics. Someday, a new relativistic quantum mechanics will be discovered in which we don't have these infinities occurring at all. It might very well be that the new quantum mechanics will have determinism in the way that Einstein wanted. This determinism will be introduced only at the expense of abandoning some other preconceptions which physicists now hold, and which it is not sensible to try to get at now. So, under these conditions, I think it is very likely, or at any rate quite possible, that in the long run Einstein will turn out to be correct, even though for the time being, physicists have to accept the Bohr probability interpretation—especially if they have examinations in front of them.[331]

It is known that Einstein rejected the quantum probabilistic interpretation with his famous statement, "God does not play dice"; a notion which is open to metaphysical assessment. However, it seems to me that the rejection of

---

328 David Bohm, *Quantum Theory* (New York: Prentice-Hall, 1951).

329 John Archibald Wheeler and Wojciech Hubert Zurek, *Quantum Theory and Measurement* (Princeton: Princeton University Press, 1983).

330 John S. Bell, *Speakable and Unspeakable in Quantum Mechanics: Collected Papers on quantum philosophy* (Cambridge: Cambridge University Press, 1988)

331 Cited in Gerald James Holton, "Einstein's Scientific Program: The Formative Years", in *Some Strangeness in the Proportion: A Centennial Symposium to Celebrate the Achievements of Albert Einstein,* ed. Harry Woolf (Reading, MA: Addison-Wesley Pub. Co., 1980), 65.

the probabilistic interpretation of quantum mechanics is caused by an intuition that occupies the minds of physicists who are accustomed to dealing with the macroscopic world on this basis. Otherwise, no strong argument is found against the indeterminism of the quantum world other than the thought experiment suggested by Einstein, by which he predicted quantum entanglement of particle pairs, which proved to be true.

In fact, the only major problem now with quantum mechanics is the issue of interpreting quantum measurements. The bizarre interpretation proposed by the Copenhagen school is not the end of the story; other interpretations might be found by which the state of indeterminism can be understood in terms better than that of a collapsing wave function, and so on. Nevertheless, indeterminism remains the correct description of the world. It is at this juncture where new developments might be initiated and where one can foresee a new scope for the extension of the quantum domain.

To finish, I would like to point out that some people might think that the laws of quantum mechanics apply only to systems on microscopic scales. This is not true. Quantum mechanics applies to all systems: galaxies, clusters of galaxies and perhaps the whole universe. However, the fact that the probabilities of macroscopic systems are very high reflects the apparent behaviour of a deterministic system. On a highly detailed scale, all systems behave along the lines of quantum mechanics, it is only that we cannot feel quantum effects on a large scale.

ABOUT THE FIRST CAUSE

It is an inherent belief in theistic Western philosophy and theology that God is the cause of the universe because everything is thought to have a cause for its existence. The argument is that, since there cannot be an infinite chain of causes, so there must be a first cause for everything. This cause is God. In fact, this assertion implicitly limits the role of a creator to the first moment. Yet, once we discover that there is no first moment for creation, we would immediately ask whether there is a place for a Creator, a question raised by Stephen Hawking on discovering that there is no t=0 moment[332] for a quantum universe, "So long as the universe had a beginning, we would suppose it had a Creator. But if the universe is completely self-contained, having no boundary

332 t = 0, the moment of the creation, called time singularity, is a moment before which nothing can be realized. But once we find that there is no t = 0, that is to say that the universe has really existed eternally therefore there can be no real t = 0 and consequently there will be no beginning for the universe. Hawking and Hartle could only show that the universe may have existed for an unlimited imaginary time before t = 0. This imaginary time however is nonphysical, as it is immeasurable, and therefore we cannot talk about a physically existing universe.

or edge, it would have neither beginning nor end; it would simply be. What place, then, for a Creator?"[333]

As we shall see, there is more than one flaw in this conclusion. The first is the assumption that the existence of God hinges on there being a beginning to the universe. The second is the assumption that the universe is self-contained. I am not sure what Hawking meant by the term "self-contained." Can any event be self-contained at all? What, then, is powering natural phenomena? Is it the inanimate laws of physics? Certainly not, since these laws have been devised by us to give an approximate, though reasonably accurate, description of the laws of nature.

As mentioned previously, the American philosopher, Adolf Grünbaum,[334] considered the Big Bang to be a pseudo-event since, at t=0, there was no time to define the start. Accordingly, he wrote, "the Big Bang does not qualify as a physical point-event of the space-time to which one would assign three spatial coordinates, and one time coordinate."[335] True as this is, it does not mean that there was no Big Bang, nor does it imply that the Big Bang was an uncaused event. Indeed, in a physical world, causal priority entails temporal priority, but this applies to the Big Bang as well. If we have to admit that all the physics that we know ceased at the moment of creation, then this means that the Big Bang might have some non-physical (I would not say metaphysical) cause, and, therefore, we may be able to admit the existence of a supernatural agent, remembering that the physics we presently know is not the only physics that humankind will ever discover.

There would be no need for a Creator if the universe had popped up spontaneously out of a vacuum, without the intervention of any other physical agent. Then, there would be no need for a first cause from the point of view of quantum indeterminacy, as it is understood now. However, it seems inconceivable that the universe, as I know it from my own research and from the work of others published in the most prestigious peer-reviewed journals, has been created completely out of nothing. Although Paul Davies admits that a highly curved spacetime should exist in order for the vacuum to be converted into energy and matter, it is unfortunate that some prominent physicists, including Stephen Hawking, Lawrence Krauss, and even Paul Davies himself, seem to overlook this fact or only touch on it, without giving it the importance it deserves. Davies, for example, considers the question of whether the spacetime

333 Stephen Hawking, *A Brief History of Time* (New York: Bantam, 1988), 141.

334 Adolf Grünbaum, "The Pseudo-Problem of Creation in Physical Cosmology", *Philosophy of Science* 56, no. 3 (1989): 373–94.

335 Adolf Grünbaum, "Creation as a Pseudo-Explanation in Current Physical Cosmology", *Erkenntinis* 35 (1994): 233–54.

warp may be regarded as a cause for the particle's appearance, as merely a matter of semantics! "The fact that the probability of particle creation depends on the strength of the spacewarp implies a sort of loose causation. The spacewarp makes the appearance of a particle more likely. Whether that is to be regarded as strictly the cause of the particle's appearance, is a matter of semantics."[336]

With this argument in mind, should we then consider that the existence of a high-energy photon, as a requirement for the creation of an electron—positron pair, is a matter of semantics? Surely not. I cannot see why the spacetime warp cannot be considered a direct cause for the first creation of matter and energy, except for the loose validity of cause in the quantum realm. Nevertheless, beyond spacetime, we do need another cause for creating an initial spacetime warp which is robust and enduring. This leaves us with something of a metaphysical dilemma.

CAUSALITY: A MODERN ISLAMIC PERSPECTIVE

According to the Qur'an, the whole universe is the grand design of Allah, who settled everything in it according to its purpose. Thus, the universe can be said to act under Divine orders and follow well-defined, comprehensible rules, which we call the laws of nature. This is the basic belief of a Muslim.

As humans, we watch events in the world and comprehend them according to the general rule of cause and effect, attributing actions to the relevant agents involved in the events. From an atheistic point of view, the availability of key material ingredients and the prevalence of favourable conditions are sufficient for a given event to happen. From a theistic point of view, the availability of these requirements is not necessarily enough to produce the event, since those requirements do not have the quality of having to make a choice or an intrinsic effect. So, the basic difference between the two viewpoints is that one considers the world to be deterministic and the other considers it to be indeterministic. However, by moving away from classical determinism, we discover that the occurrence of an event, or process, takes much more than the required conditions and goes beyond the power of the individual entities that contribute to the event. This motivates one to think of an agent who dominates the whole universe with a plan, and also of a power that makes it possible to have an ordered world, which is ruled according to pre-set laws that we try to recognise by observing the world's phenomena. We also need to explain how the ingredients of the alleged cause are acting according to their own nature, in order to justify the outcome.

336 Paul Davies, *God and the New Physics*, London: Penguin Books, 1990, 35.

Certainly, the concept of causality and the understanding of causal relationships is one of the vital subjects that need to be revised and set in a new Islamic perspective. In recent years, a few Muslim scholars have tried to reform the understanding of causality and causal relationships in modern Islamic thought. However, none of them is known to have proposed a self-contained theory of causality that could mitigate the conflict between contemporary scientific facts and the understanding of religious beliefs.

In an article entitled, "Quantum Theory, Causality and Islamic Thought",[337] Mehdi Golshani presents what he perceives to be an Islamic outlook on causality. Golshani discusses this concept in the context of modern physics, more specifically in the context of quantum theory. Admittedly, he agrees that quantum theory has abolished classical causality and causal determinism, but he is inclined to support the effectiveness of classical causality in nature. For this reason, Golshani has tried to present an Islamic view for causality with an affirmative voice, using a few verses of the Qur'an, which refer to law and order in nature and to the possibility of some secondary causes (which some verses indirectly point to). Golshani also puts forward some prospects for replacing quantum theory with an alternative which might restore causality and causal determinism. This kind of argument is no more than wishful thinking. There is no point in dreaming of restoring determinism, since experiments are being conducted daily which confirm the predictions of quantum mechanics. Here, I cannot say more than what John Wheeler expressed in the quote above, which Golshani himself has quoted too. New sub-quantum physics might explain indeterminism, but it certainly will not be able to restore determinism, since indeterminism is a profound property of phenomenon in the quantum realm.

In an article entitled, "Causality and Divine Action: The Islamic Perspective", Mohammad Hashim Kamali[338] presented a view that endorses the affirmation of causality. Again, this author lays stress upon the claim, without providing enough evidence, that Muslims and the Islamic attitude would support the existence of causality and causal determinism in the world. Nevertheless, Kamali recognised that, "The Qur'anic evidence on causality and its related themes, free will and determinism is not decisive on any particular position or view that has been taken over these issues. This also means that the Qur'anic evidence on causality is two sided and remains open to interpretation. Numerous passages are found in the Qur'an where God, most High, identifies Himself as the only cause and originator of things."

337 Mehdi Golshani, "Quantum Theory, Causality and Islamic Thought", in *The Routledge Companion to Science and Religion*, ed. James W. Haag, Gregory. R. Peterson, and Michael L. Spezio (London: Routledge, 2012), 179.

338 Mohammad Hashim Kamali, "Causality and Divine Action: The Islamic Perspective", Ghazali.org, accessed 26 June 2015, <http://www.ghazali.org/articles/kamali.htm>

CAUSALITY IN THE NEW *KALĀM*

In the following section, I will propose a reformulation of the concept of causality within the new *Daqīq* al-*kalām* approach. I will take the truth of Islam as a prime input and consider the facts of science and its methodology for investigating the world as the proper approach for obtaining knowledge about it. This approach enables us to verify our understanding of religion. This new reformulation of the concept is deeply rooted in the original Islamic understanding of the world. The views and explanation given here will utilise the results obtained from Chapter Eight that gives a complete interpretation of quantum mechanics and quantum measurements.

Historically, the concept of causality was always under the influence of the basic doctrine of Islamic belief, which asserts that Allah is the only cause of everything that happens in the world. On the other hand, causal relationships suffered from confusion and much distortion, being perplexed by the influence of observations which assert the action of the laws of nature, its reliability, and efficiency. At the same time, we have to be faithful to what Allah has clearly stated as the truth of Islam in the Qur'an. It would not be a proper approach in the reconciliation of religion with science to re-adjust our interpretations of the Holy text whenever a new scientific discovery is found. This suggestion, which was originally proposed by Ibn Rushd,[339] is certainly nugatory since we now know that some of the so-called facts of science change continuously with the development of the means and techniques used.

As I have mentioned in the section concerning the phenomenology of a cause, the development of science brings in new concepts which suggest new causes for phenomena. For example, Einstein's theory of gravity replaced the Newtonian concept of a gravitational field with the notion of spacetime curvature. This produced a new vision for understanding the cause of gravity instead of attributing it to the masses of interacting objects; general relativity attributes the gravitational interaction to the curvature of spacetime in which those masses exist. On the other hand, the concept of action-at-a-distance, which implies infinite speed for gravitational interaction, was demolished by Einstein's Theory of Relativity, whereupon the gravitational effect was found to propagate through space at the speed of light. Many other such examples can be found in chemistry, biology, and medicine. We all know how many times theories of disease and the diagnosis of causes in the medical profession change. This makes one conclude that our designation of causes is mostly tentative and will never be covered with absolute certainty. Therefore, the development of scientific theories sometimes produces drastic changes in identifying the causes

339 Ibn Rushd, *Fasl*, p.32.

and, even, the subsequent effects. However, causal relationships are an established fact of the world's phenomena; no phenomenon in the world can occur without a formal cause that is assigned in association with certain parameters which are related to the innate properties of space, time, and matter.

However, causes cannot be directly attributed to God since we do not know the Divine mechanism through which His action is taken to establish the effect. On the other hand, we cannot consider the laws of physics, which we identify through studying phenomena, to represent a Divine mechanism, since these laws are the products of our own minds and not the mind of God.

The assumption of re-creation can play a fundamental role in formulating the concept of causality. Using this assumption, it is possible to derive many consequences that agree with, but do not depend on, the quantum description of the world by which causal determinism was abolished. At the same time, it provides a vivid understanding of the role of the Divine in sustaining the world. The re-creation assumption implies that some uncertainty is inevitable in any simultaneous measurement of the so-called incompatible observables, such as position and momentum, time and energy.[340] Re-creation is affected by the innate properties of an entity and, consequently, any change that takes place due to the interaction between two such entities is to be viewed as a new state that resulted from re-creating the respective properties, which duly changes the character of those entities. In order to see how re-creation can preserve causal relations, we should take into consideration two facts:

1. Change is a basic character of the world. Taken on a microscopic scale, the persistence of change is clear and vivid; even a vacuum is known to be a boiling pot of virtual states. Taken on the macroscopic level, however, change is hardly noticeable.
2. The indeterminism of the world is clearly exhibited within the microscopic world, whereas events of the macroscopic world are *almost* all deterministic.

An assumption of re-creation in the innate properties of entities allows for an explanation of indeterministic causal relationships. These relationships are understood to be the means through which events in nature are related to each other. Obviously, not all events are related to one another; nevertheless the re-creation postulate allows for the whole world to be interconnected. This is also part of the holistic entanglement of the universe that is implied by some interpretations of quantum physics.

340 For a detailed technical consideration, see Mohammed Basil Altaie, "Re-Creation: A Possible Interpretation of Quantum Indeterminism", in *Matter and Meaning*, ed. Michel Fuller (Newcastle Upon Tyne: Cambridge Scholar Publishing, 2010), 21; available at arXive: quant-ph/0907.3419v1.

The mechanism of re-creation is as follows: primarily, we assume that the innate properties of entities allows for their interaction capabilities. Since no proof is available for deterministic causality, such interactions should be considered contingent, but not necessary, and always dependent on the efficacy of innate properties. Under re-creation, the efficacy of innate properties changes (maybe chaotically). The frequency by which these changes occur is directly proportional to some physical parameters related to the system; namely the energy of the system. High-energy systems acquire high frequencies of re-creation, while low-energy systems acquire low frequencies. Every time a physical (observable) property changes due to re-creation, the efficacy related to that property will change, assuming a new value and, consequently, the causal event will assume a new status. This will result in maintaining the indeterministic causal relationship between parts of the world. Note that with this vision the explanation of causality no longer follows an apologetic approach or an occasionalist understanding; rather it is a scientific vision that is based on mathematical description and proof.

The concept of causality being associated with Allah in Islamic *kalām* is fundamentally tied to the principle of re-creation, one of the general principles adopted in an attempt to gain an understanding of the world. This principle (or assumption) reproduces the indeterministic character of the world. The affiliation of causality with the agency responsible for re-creation makes for a logical conclusion. This association does not mean that the world is being run miraculously, since it was originally designed to respect certain laws—that is the laws of nature—which are firm and immutable. According to these laws, the occurrence of an event is probabilistic and would result in one of the contingent states that are permissible by these laws. Indeterminism, then, is reflected objectively in the choice of the outcome, which cannot be accounted for arbitrarily. The observation that evolution of the world is progressing towards higher and better states is an indication that the choices made from that probabilistic outcome are deliberate and purposeful.

However, causal relations are something else. Whereas the prime cause for every event is Allah, and whereas Allah is the initiator who fuels the laws of nature, causal relationships are deemed to be a part of the original creation by which events and processes transpire through the course of well-defined algorithms and laws which exploit the innate properties of things (which should be understood as being purposely allocated by the Creator), expressing the indeterministic causal relationships in the form of the laws of nature. The innate properties are real and objectively exist but are continuously re-created. Such a continuous re-creation allows for change to take place once any object acts on any other object, thereby influencing it with its properties. That is to say that re-creation allows for the change in the innate properties to take place through the interaction between objects. Otherwise, if innate properties were immutable, it would be hard to see how it could change at all, even with the influence of one object over the other. This is why Muslims did not accept the concept of causality as proposed by philosophers, or the role that they gave to the nature of things in causal relationships.

Within this view, there is no denial of causality. Causal relationships are respected, but at the same time, the Divine will and choice is fully respected too. There is also no place for causal determinism as there are two agencies at play: God and the laws of nature. On the other hand, although God's role is understood to be supreme, He is not a tyrant puppeteer. This understanding of causality and causal relationships provides us with the necessary explanation to enable us to comprehend the role of blind laws that act efficiently in the world, and to envisage how the world is progressing on an evolutionary level in the presence of those conflicting blind laws of nature.

Chapter Eleven

Astronomy and Cosmology

The Mutakallimūn dealt with questions concerning astronomy and cosmology as part of their philosophical questions and interest. Most of the early Mutakallimūn of the 9th and 10th centuries believed that the Earth was flat and stationary. Ibn Ḥazm and al-Ghazālī acknowledged that it was spherical but did not realise that it moved. Many of the later Mutakallimūn, that is, after the 12th century, realised that the Earth was spherical but, again, did not predict that it was mobile.

However, most of the early Mutakallimūn did not accept the Greek theory regarding the four elements and classification, or their descriptions of the upper and lower worlds. They also rejected astrology on religious grounds since Prophet Muhammad (peace be upon him) had warned Muslims of dealing with astrology for the sake of knowing the future or telling fortunes. Al-Bāqillānī, in particular, rejected the notion of four elements and four natures, and also negated the influence of the planets and the signs of the zodiac on people or anything on Earth. As they noted that everything falls downward, they found the need to explain how the Earth could stay static in space. Some of those who believed that the Earth was still, tried to explain it by proposing the balancing of forces from below and from above. Al-Ashʿari details some of these suggestions.[341] He also reports Abū al-Hudhayl as saying that it stands still without any assistance, as God is able to support it as such.[342]

In most cases, knowledge about the Earth and sky that was available to Muslims around the end of the 10th century was sourced from the contents of Ptolemy's *al-Magestic*, which was translated at the time of the second Abbasid Caliph Abū Ja'far al-Mansoor (712-775). However, they disagreed with the Aristotelian philosophical classification of the world and the heavens. Al-Ghazālī and later, Ibn Rushd, were among those theologians to have acquired, perhaps, the best astronomical information of their time. This is clear from the examples and the discussions of the *Incoherence* (*Tahafūt*).

In this chapter, I will present the discussion of Abū Hamid al-Ghazālī on two astronomical topics: the first question is related to the size of the world and whether beyond it there could be a full space or a void. The second question concerns the aging of the Sun and whether it could corrupt and degenerate into other elements.

The first question focuses on the relativity of designating time (the before and the after), which al-Ghazālī likens to the relativity of space (the above and the below), and whether time had a beginning or not. In this discussion, al-Ghazālī settles on the decision that time must have a beginning which accompanied space and matter. Moreover, al-Ghazālī deduces that time should be treated on an equal footing, as space in that time is a dimension, the same as space being a dimension. Consequently, time has no absolute reference, except with respect to the moment of creation. In order to establish this analogy, al-Ghazālī went on to question the theoretical possibility that the universe could have been created larger or smaller than it is. If such a possibility is allowed, then the question arises about whether the universe has an exterior into which it can be extended. This was a challenging question for the philosophers, who denied such a possibility, always asserting the finiteness of the universe within a fixed size and shape. Hence, they had no alternative but to admit that beyond the universe there could be neither a body nor a void. This would mean that the volume of the universe is all which can be recognised to exist. Accordingly, and since time is associated with space and matter, no recognition of time would be possible unless the universe existed. Hence, the question of a time existing before the creation of the universe is deemed to be meaningless.

The second question queries Galen's (and the Aristotelian philosophy) doctrine that the Sun is an ethereal body that does not weather or decay. This was presented in the *Tahāfut al-falāsifah* in the context of discussing a post-eternity of the world, time and motion. In this respect, al-Ghazālī did not

341 Al-Ash'ari, *Maqālāt al-Islāmiyyīn*, 2:249.
342 Ibid. 412.

deny the possibility of a post-eternity of the world on a rational basis, but he suggested that this could only be denied on a religious basis. Al-Ghazālī refuted the argument which had been made to affirm the Sun's eternity, thereby, rejecting any notion of its post-eternity. That particular argument had taken the observational fact that the Sun looked the same, without withering, for a very long period of time as evidence for its eternity. Al-Ghazālī showed that this argument was flawed, since it was based on observations that were inadequate for detecting small losses of mass from the body of the Sun that might be taking place at a very slow rate. Al-Ghazālī concludes that this doctrine is unfounded, as the Sun might be in decay already; a striking conclusion for his era.

THE INCOHERENCE OF THE PHILOSOPHERS

In the history of Islamic philosophical thought, the arguments presented by al-Ghazālī and Ibn Rushd on issues of natural philosophy stand as a remarkable monument that reflect the high intellectual level and standard of discourse that existed in their times. Such a splendid debate is a good original resource for identifying the intellectual themes at the time that al-Ghazālī wrote his book, *Tahāfut al-falāsifah* (*The Incoherence of the Philosophers*), in which he endeavours to refute the philosophical approach for comprehending the relationship between God and the world, and to present an alternative view from that of the Islamic perspective. It might be useful here to present the basic motivations for al-Ghazālī to write his book; for this purpose, I will use a sizeable number of quotations from his book.

He thought that following the philosophical approach would corrupt religion:

> [A] group who, believing themselves in possession of a distinctiveness from companion and peer by virtue of a superior quick wit and intelligence, have rejected the Islamic duties regarding acts of worship, disdained religious rites pertaining to the offices of prayer and the avoidance of prohibited things, belittled the devotions and ordinances prescribed by the divine Law, not halting in the face of its prohibitions and restrictions.[343]

Al-Ghazālī also wrote that:

> The source of their unbelief is their hearing high-sounding names such as Socrates, Hippocrates, Plato, Aristotle, and their likes, and the exaggeration and mis-guidedness of groups of their followers in describing their minds, the excellence of their principles, the exactitude of their geometrical, logical, natural and metaphysical sciences, and in [describing these as] being alone by reason of excessive intelligence and acumen [capable] of

343 Al-Ghazālī, Incoherence, 1-2.

> extracting these hidden things; [also hearing] what [these followers] say about [their masters, namely] that, concurrent with the sobriety of their intellect and the abundance of their merit, is their denial of revealed laws and religious confessions, and their rejection of the details of religious and sectarian [teaching], believing them to be man-made laws and embellished tricks.[344]

Furthermore, Al-Ghazālī wrote:

> In refutation of the ancient philosophers, to show the incoherence of their belief and the contradiction of their word in matters relating to metaphysics; to uncover the dangers of their doctrine and its shortcomings, which in truth ascertainable are objects of laughter for the rational and a lesson for the intelligent, I mean the kinds of diverse beliefs and opinions they particularly hold, that set them aside from the populace and the common run of men.[345]

In his treatise, al-Ghazālī mostly adopted the views and approach of *kalām*. It was his intention to expose the failure of the philosophical argument in proposing a truly reconciliatory view of God and the world to which Muslims could subscribe without infringing their beliefs. He brought the conflict between Islamic *kalām* and philosophy to a head by undertaking a refutation of twenty philosophical doctrines, out of which seventeen were condemned as being heretical innovations and three as being totally opposed to Islamic belief. In part, al-Ghazālī targeted two prominent Muslim philosophers, al-Fārābī and Ibn Sīnā, for their views about the emanation of the world and the resurrection of bodies. The philosophers he condemned were not atheists and, in fact, their philosophies rested upon the affirmation of God and the recognition that everything in existence emanates as the necessary consequence of Divine essence. Nevertheless, al-Ghazālī saw this as meaning that God had produced the world through necessity, in the same way that an inanimate object, such as the Sun was said to produce light by its very nature. For him, the views presented by these philosophers meant the denial of the Divine attributes of life, will, power and knowledge. Al-Ghazālī maintained that, by denying these attributes, the God of the philosophers was not the God of the Qur'an.

Despite the fact that his *Tahāfut al-falāsifah* brought the conflict between philosophy and more traditional Islamic beliefs to the fore, al-Ghazālī compiled this treatise in order to explain some important philosophical arguments. Owing to its intellectual calibre, his book marks a high point in the history of medieval Islamic thought. Although its motivation was religious, as shown

344 Ibid., 2.
345 Ibid., 3.

above by his own words, it made its case through closely argued criticisms that were, ultimately, philosophical. It is important to note that al-Ghazālī based his critiques of those philosophical arguments, which were related to natural philosophy, on a consistent body of thought which embodied the principles of *Daqīq* al-*kalām*. I have presented and studied these in Part One of this book, in a most efficient manner.

This, as we will see below, was not an isolated argument, but rather a full range of discussions which encompassed a detailed comprehension of fundamental concepts such as space, time, causality, the laws of nature and several others. To a large extent, one can say, without any equivocation, that al-Ghazālī affirmed the Ashʿari theory of causality. For him, Divine power is pervasive, and is the direct cause of each and every created existent, as well as each and every temporal event. The basic principle he adopted is that inanimate objects have no causal power, a view which he asserted in other works, such as *al-Iqtiṣād fi al-i'tiqād*.

THE INCOHERENCE OF THE INCOHERENCE

Ibn Rushd, a philosopher in Islamic Spain, adopted the views of Aristotle and provided great service to Greek philosophy by explaining Aristotle's legacy. He found the work of al-Ghazālī to be a sort of forgery of philosophy. Accordingly, he devoted his work *Tahāfut al-Tahāfut* (*The Incoherence of the Incoherence*) to trying to refute al-Ghazālī's arguments, following him paragraph by paragraph, and mostly reinstating the position of Aristotle on those questions. I will consider quotations from this book in due time when considering the response of Ibn Rushd to al-Ghazālī.

In this debate al-Ghazālī was speaking on behalf of the Mutakallimūn while Ibn Rushd was defending the philosophers. Comparing the two arguments and views might enable the reader to appreciate the way of thinking of both scholars and assess the methodology of deduction they followed.

Both problems were reconsidered by Ibn Rushd in his defence of the philosophers' arguments. Ibn Rushd had fully adopted Aristotelian cosmology and, accordingly, tried to show that the universe could not be larger or smaller than its given size, because such a possibility would change the universal order. He also supported Galen's position concerning the eternity of the Sun. In his response to al-Ghazālī's work, Ibn Rushd was mostly apologetic and did not provide enough evidence, other than what was available in Aristotle's metaphysics. However, at certain points in the discussion, one cannot deny that Ibn Rushd made certain notes and conclusions that reflected his own genius.

The presentation of the two problems here is intended to provide important historical examples of how the views of *kalām* and those of the philosophers constituted two rival theories in respect of natural events. In concluding this chapter, I will try to assess the views that al-Ghazālī and Ibn Rushd presented on these two problems from the current scientific standpoint, which is based on the discoveries of modern astrophysics and cosmology.

GREEK SCIENCES IN ISLAM

The creation of the firmament (including all the celestial objects) and the design and development of the cosmos are amongst the main considerations of many verses of the Qur'an. The primary goal in presenting this issue, it seems, is to turn people's attention to the signs that demonstrate the glory of the Creator who designed the cosmos, and to appreciate the need for such a Creator and designer. The signs mentioned in the Qur'an encouraged Muslims to contemplate the cosmos and try to understand God's action in the world through pure reason, according to the philosophical trend. On the other hand, the transmission of Greek astronomy contributed positively to the scientific movement in the Islamic world, to the extent that Muslim astronomers were able to develop their own new techniques for astronomical observations and to criticise Ptolemy's geocentric system, as well as developing an alternative astronomical system to replace it.[346]

Most Muslim scientists and philosophers were overwhelmed by Greek sciences and philosophy such that they could not circumvent the main propositions of the Greeks with respect to their views of the cosmos. This applied to philosophical trends as much as to scientific trends in Islam. Greek philosophers, such as Plato and Aristotle—"men of wisdom", as they were called in the Islamic media—were highly respected by many Muslim philosophers, especially al-Fārābī, Ibn Sīnā, and Ibn Rushd. Some Muslim philosophers admired Plato and Aristotle to the extent that they considered them to be gifted, wise men, who were comparable to prophets; a status which would make it nearly impossible to breach their doctrines. For this reason, it is not surprising to learn that the genuine contributions by Muslim philosophers to the philosophical achievements were quite modest, and that most of their works echoed the original Greek views, despite their endeavours towards reconciling Islamic belief with philosophy.

On the other hand, Muslim rational theologians who were rehearsing an independent style of thought based on the Qur'an, were less affected by the philosophical considerations of the Greeks. This strongly applied to the pioneering practitioners of *kalām* and the Mutakallimūn, who were active during the 8th and 9th centuries. This is why we see that the Mutakallimūn were able to reflect the true values and originality of the Islamic creed.

---

346 Noel M. Swerdlow and Otto E. Neugebauer, *Mathematical Astronomy in Copernicus's "De revolutionibus"* (New York: Springer-Verlag, 1984).

Apart from some minute details, the picture that Muslims had about the firmament was almost the same as the one that was adopted and developed by the Greeks, in which the five planets, the Sun, the Moon, and the fixed stars were thought to be associated with concentric celestial spheres, at the centre of which the Earth resides. This is the so-called "geocentric model" of the planetary system. In its most sophisticated form, this model, as mentioned above, was devised by Ptolemy and was adopted to calculate the positions of the celestial objects for more than 1,200 years. The fact that the calculations did not accurately fit with observations made, required many readjustments of the model through the well-known epicycle assumption, by which astronomers suggested new values for the parameters, whenever they found that their calculations did not fit the actual observations.

THE FIRMAMENT IN THE QUR'AN

The terms heaven (singular) and heavens (plural) have been mentioned in the Qur'an three hundred and ten times. In so many verses, these terms describe Divine providence and God's care of mankind. The Qur'an stresses that Allah has created the firmament and that He is developing it. However, the concept of the firmament does not have a definitive meaning in the Qur'an. It could mean the clouds, the atmosphere, the open sky, the solar system, the Orion arm of the galaxy to which we belong, the galaxy, the whole universe, or even other universes that are beyond our comprehension. The meaning of the word firmament is so widely presented that one cannot help but treat every mention of it within its given context.[347] What is of interest at this point is the way that the Qur'an presents the development of the firmament. For example in Surah Al-Anbya:

> *Have not those who disbelieve known that the heavens and the earth were joined together as one united piece, then We parted them? And We have made from water every living thing. Will they not then believe? (21:30)*

Despite some contextual similarities, this description for the creation of the firmament is different from what is written in the Old Testament. Here, we understand that the heavens and the earth were one entity and they were then separated. The heavens were not born from the surface of water, as it is stated in the Old Testament. Nevertheless, creation has originated from water in the sense that water is the basic composition of living creatures. This provides a more detailed and accurate picture of the event of creation.

---

347 Mohammed Basil Altaie and M.K. al-Zuʿbī, "The Concept of Heaven and Heavens in the Qurʾān and Modern Astronomy", *Jordanian Journal of Islamic Studies* 4, no. 3 (2008): 223–49.

Moreover, it seems that this event of creating the firmament is continuous; in Surah Adh-Dhariyat:

> *We constructed the firmament with our hands, and we will continue to extend it. (51:47)*

This clear statement shows the development of the firmament, after its creation, to be one of continuous expansion. Such an expansion might entail the continuous creation of space, but it is not clear whether this creation is accompanied by the creation of matter and energy, or whether it is only space that is expanding. Neither does the Qur'an specify where expansion is taking place, whether it is within a larger volume of space or whether it is expanding from within. The word "extending" is different from the word "expanding", which has been used by some translators of the Qur'an. To expand may mean to increase the space between the constituents of a structure, say a city, without adding any new buildings. But to extend a city implies the adding of new buildings and an area to it. In the above verse, the sentence in the Qur'an indicates that the firmament constitutes a block that is being extended by the addition of new structures and spaces. So, one may say that space is being expanded possibly with the addition of new matter.

However, according to the Qur'an, such an expansion will not go on forever. It has to stop one day after which it will start to contract, taking the universe back to where it started; in Surah Al-Anbya we read:

> *And (remember) the Day when We shall roll up the heavens like a scroll rolled up for letters, as We began the first creation, We shall repeat it, (it is) a promise binding upon Us. Truly, We shall do it. (21:104)*

This clearly indicates that the firmament is going to contract and revert to the state of its first creation. In modern cosmology, this is called the "Big Crunch." The rolling up of the scroll is an interesting image which indicates a flat universe. Arguably, recent discoveries of the accelerating universe may indicate that the universe will go on expanding forever. Nevertheless, this fate might be challenged by introducing other factors which affect the status of the universe, such as the so-called "cosmological constant," while some studies allow for the possibility of a collapsing flat universe.

Ikhwān al-Ṣafā' (the Brothers of Purity) was a secret group of the Ismāilī sect, which adopted the Aristotelian model of the heavens and tried to integrate it with Islamic belief through interpretations of the relevant verses of the Qur'an. They imagined seven ethereal spheres holding seven celestial bodies as being the seven heavens and the seven earths that the Qur'an mentioned. So, for them, the world was no more than these objects surrounded by a sphere containing the fixed stars and encircled by the Atlas orbit, which they interpreted as the throne of God. However, they did not tackle the question of the expanding or extending world mentioned in Surat al-Dhāriyāt. For this reason, we can say that their Aristotelian view was not of much help in covering the stipulations of the Qur'an concerning the development and fate of the universe.

On this point, al-Ghazālī did not use a religious argument when asserting the possibility of an expanding universe; although, he may have been motivated by the stipulations of the Qur'an concerning such a possibility. Contrary to the mainstream philosophical thinking at that time, which considered the universe to be static and eternal, al-Ghazālī believed that the world was temporal, being created out of nothing, *ex nihilo*, and that its creation marked the beginning of both space and time. This understanding was actually borrowed from the Mutakallimūn, who, earlier in Islamic thought, had devised a theory of creation by which the universe came into being through the sheer will of Allah. In our present time, William Craig has made some serious efforts to elaborate on the *Kalām* Cosmological Argument, which stipulates that the universe must have a cause for its existence since it has a beginning.[348]

Al-Ghazālī on the Size of the Universe

In the first discussion of his *Tahāfut al-falāsifah*, al-Ghazālī discussed the problem of the temporality and the eternity of the world. His strategy was based on defying what he considered to be the strongest arguments of the philosophers in claiming that the world should be eternal, raising some challenging questions for the philosophers through discussing their views and exposing inconsistencies in their arguments. In this context, al-Ghazālī presented a very deep and thoughtful discussion of space and time, defending the necessity to recognise the fact that space and time allocations should not be taken as absolute, but should always be considered in reference to a given point in space or time. This was a very advanced comprehension of a topic that might well be considered a problem for the modern science of the 20th century.

Al-Ghazālī used the terms "spatial dimension" and "time dimension."[349] He refused to acknowledge both the notion of a space that goes beyond the world and the existence of time before the creation of the world.[350] It was through this comparison between space and time that he introduced the

348 Craig, *Kalām Cosmological Argument*.

349 Considering time as a dimension on equal footing with space is a fundamental concept that was introduced in modern times by Albert Einstein through his theory of relativity. The three known spatial dimensions were integrated with the time dimension to form the spacetime continuum.

350 No doubt, the notion that time only existed along with the creation of the world was part of the propositions of St. Augustine in his *Confessions*.

question about the size of the world, allowing for the possibility that the world could have been created larger or smaller than it is. With his sophisticated concept of space and time, and his realisation of the analogy between space and time, al-Ghazālī contested the philosophical claim that an infinite extension of time should have existed before the creation of the world. The most important argument which was placed in this context was the notion that both space and time existed only after the creation of the world, a concept that has been established only by the modern theories of cosmology.

One of the interests of al-Ghazālī concerned the size of the universe, where he questioned whether the universe could be larger or smaller than it is. This he did in order to challenge the philosophers, trying to force them to admit one thing or another in their views concerning the existence of time before the creation of the universe. The philosophers used to argue that, if the universe were not eternal but had been created in time with a well-defined beginning, then why did the Creator wait so long before creating it? Obviously, this question implicitly assumes that the Creator lives in time.

Al-Ghazālī first questioned the philosophers over whether God could have created the world larger than its known size, "Did it lie within God's power to create the highest heaven greater in thickness by one cubit than the one He had created?."[351] Then he commented, "If they say, 'No,' this would be [the attribution to Him of] impotence. If they say, 'Yes,' then [it follows that God could have created it] greater by two cubits, three cubits, and so on, ascending ad infinitum."[352] Consequently, al-Ghazālī concluded that if the answer was 'Yes', then this would imply the affirmation of a space beyond the world that has a measure and quantity, since that which is greater by two cubits does not occupy the equivalent space as the one greater by one cubit. Accordingly, he said, "Then, beyond the world there is quantity, requiring thus that which is quantified—namely, either body or the void. Hence, beyond the world there is either void or filled space."[353]

By putting forward this point, al-Ghazālī posited a fundamental paradox that the philosophers were required to solve. The paradox had two possible rejoinders. They could have said that beyond the world there is a void into which it could be expanded. But the existence of such a void went against the

351 Al-Ghazālī, *Incoherence of the Philosophers*, 37. Clearly in his question al-Ghazālī is talking here about God's absolute power, arguing for the possibility of having a universe larger or smaller that its known size. The argument is equally the same if he had said it would have been possible that the universe was created larger or smaller by one or more cubits. The allowance of such a possibility confirms that he was convinced by such a possibility, though he did not provide a natural reason for it.

352 Ibid.

353 Ibid. 38.

doctrines of the philosophers, who refused the existence of voids anywhere in the world. Alternatively, they could have said that beyond the world there is a matter-filled space. In this case, there would be no reason why such a filled space should not be part of the world itself, since it would then be no more than an extension of it.

Additionally, al-Ghazālī posed the question of whether God is able to create the world's sphere smaller by one cubit, then by two. Accordingly, if one could accept that the measure of the world is reducible in size then, al-Ghazālī suggested, this would imply that the void which is left when we reduce the size of the world is measurable, while still being nothing. The other side of the paradox was to challenge the philosophers about the limit of God's authority with respect to creating and sustaining the world; a challenge that Muslim philosophers certainly would not have been able to stand.

In fact, the aim behind these questions regarding the size of the world was tactical rather than strategic. Al-Ghazālī had no intention of showing that the universe could be expanded or contracted, he intended to show, only, that we must consider the temporal designations in respect of the *before* and the *after* on an equal footing with the spatial assignments of the *above* and the *below*. That is to say the temporal assignments of events should be allocated with respect to a given reference rather than being absolute. Therefore, al-Ghazālī's argument served a dual purpose: one, by which he intended to show that there is no basic natural objection to having a universe larger or smaller than the existing one, and the other, that such a possibility would certainly reassure the conceptual integrity of space and time. Consequently, he made an effort to use these results to contradict the claim that a temporal world necessitates the existence of a time duration *before* creation had taken place. For this reason, it could be said that al-Ghazālī would be quite happy with Adolf Grünbaum,[354] who recently argued that the moment of creation does not qualify as a physical event, since there was no physical moment *before* the initial moment of the Big Bang.

Indeed, according to al-Ghazālī, the creation of the world did not happen *in* time but *with* time. For this reason, it is legitimate to suggest that there is no well-defined moment of creation, since real time only started *with* that moment. This would indeed be quite consistent with an earlier discourse of al-Ghazālī's:

> Similarly, if we are asked: does the world have a "before"? We answer: If by this it is meant does the world's existence have a beginning, that is, a limit in which it began, then the world has a "before" in this sense, just as the world has an "outside" using the interpretation that this is its exposed

354 Grünbaum, "Creation as a Pseudo-Explanation."

> limit and surface end. If you mean by it anything else, then the world has no "before," just as when one means by "outside the world" [something] other than its surface, then one would say: there is no exterior to the world. Should you say that a beginning of an existence that has no "before" is incomprehensible, it would then be said: a finite bodily existence that has no outside is incomprehensible: If you say that its "outside" is its surface with which it terminates, [and] nothing more, we will say that its "before" is the beginning of its existence which is its limit, [and] nothing more.[355]

So, it is here that the moment of creation is considered unique, in that it has no similarity with any other subsequent moment. To confirm this, al-Ghazālī further emphasised the premise that God is timeless and, therefore, the question of what God was doing before the creation of the universe becomes meaningless; a position similar to that put forward by St. Augustine. Al-Ghazālī's genius is clear in his rhetoric, which makes one admire the consistency of his thinking and the boldness of his proposal that time should be treated on an equal footing with space. It might be said that the visualisation of time existing alongside the creation of the universe had already been mentioned by St. Augustine. This is true and it might be that al-Ghazālī adopted this visualisation. However, it is important to note that he extended the analogy of space and time into a realm well beyond that of St. Augustine by recognising that the universe has no exterior, and that the temporal extension is a dimension that has to be treated on an equal footing with spatial dimensions. This is a concept that we acknowledged in our modern age only after Albert Einstein's discovery of relativity.

Ibn Rushd's Response

In his book, *Tahāfut al-Tahāfut (The Incoherence of the Incoherence)*, Ibn Rushd tried to refute the claims of al-Ghazālī by criticising his arguments and presenting counter-arguments. As far as the size of the universe is concerned, Ibn Rushd denied that the philosophers had said that God could not change the size of the universe, and rejected the accusation that their position on this matter implied that God is impotent, "This is the answer to the objection of the Ashʿaris that to admit that God could not have made the world bigger or smaller is to charge Him with impotence, but they have thereby compromised themselves, for impotence is not inability to do the impossible, but inability to do what can be done."[356] Clearly, to say that impotence is not the inability to do the impossible but the inability to do what can be done is true with respect to human acts, but not to Divine acts, because we are not sure whether anything is impossible for God. Ibn Rushd confirmed this attitude by saying:

355 Al-Ghazālī, *Incoherence*, 35.
356 Ibn Rushd, *Tahafut al-Tahafut.*

> This consequence is true against the theory which regards an infinite increase in the size of the world as possible, for it follows from this theory that finite things proceed from God which are preceded by infinite quantitative possibilities. And if this is [an] allowed for possibility in space, it must also be allowed with regard to the possibility in time, and we should have time limited in both directions, although it would be preceded by infinite temporal possibilities.[357]

He then concluded:

> The answer is, however, that to imagine the world to be bigger or smaller does not conform to truth but is impossible. But the impossibility of this does not imply that to imagine the possibility of a world before this world is to imagine an impossibility, except in the case where the nature of the possible was already realised and there existed before the existence of the world only two natures: the nature of the necessary and the nature of the impossible. But it is evident that the judgement of reason concerning the being of these three natures is eternal, like its judgement concerning the necessary and the impossible.[358]

This means that it is not at all possible for the size of the universe to be smaller or larger than it is but is something which falls between being either necessary or impossible. With this digression, Ibn Rushd shifted the argument away from the arena of metaphysics to the arena of physics. Using such a designation, Ibn Rushd thought he could refute al-Ghazālī's conclusions and win the argument. From his point of view, it is impossible for the universe to be larger or smaller than its natural size, since the specified size of the universe is a necessity. As for the designations, necessity and impossibility, it is clear that Ibn Rushd was adopting the naturalistic dogma, which assumes that whatever happens in the world has to be affected by purely natural causes and that it should take place in harmony with the laws of nature. However, this can be validated only if we have a complete knowledge of the laws of nature, but since we know that our knowledge of these is incomplete, it would be humbler to allow for the possibility of the event happening, rather than deny it. This is, in fact, the contemporary approach adopted by modern scientists and according to which, new discoveries are made.

Ibn Rushd further embraced his denial of a possibility for the universe to be larger or smaller than its known size, by substantiating his views with more arguments which stemmed, perhaps, from his inability to visualise time on an equal footing with space. Thus, he was unable to accept the notion of spacetime integrity and the absence of absolute space and absolute time; such

357 Ibid. 65.
358 Ibid.

points which were essential to the argument used by al-Ghazālī. In fact, Ibn Rushd suggested that if the universe were allowed to expand, then there was no reason why it should not do so forever, "Therefore, he who believes in the temporal creation of the world and affirms that all body is in space, is bound to admit that before the creation of the world there was [a] space, either occupied by body, in which the production of the world could occur, or empty, for it is necessary that space should precede what is produced."[359] Again, it is clear that Ibn Rushd missed the point made by al-Ghazālī that space itself was non-existent before the creation of the world. This is because he thought of space and time as two independent entities. From al-Ghazālī's point of view, the existence of an empty space into which the universe could be extended would be unnecessary as space was born along with the creation of the universe. The same argument applies to time, since space and time are integrated, according to the understanding of the Mutakallimūn—as stipulated in the fifth principle discussed in Chapter Seven.

Clearly, al-Ghazālī had allowed for two possibilities for the universe to be larger or smaller than it is. He could foresee no rational reason to prevent such a possibility. It might be true that his attitude stemmed from his unwavering belief in the unlimited power of Allah to do whatever was contingent. On the other hand, Ibn Rushd based his argument on the Aristotelian proposition that the size of the universe is fixed, and that no other possibility is viable. His conclusion that if the universe were "allowed" to be bigger there would be nothing to stop it from expanding further, is unacceptable since this would lead to an infinite universe, once we assume that it had no beginning, a result which would be in contradiction with the Aristotelian doctrine of a finite universe. Aristotle argued that the universe is spherical and finite. Spherical, because that is the most perfect shape; finite, because it has a centre, namely, the centre of the earth, and a body with a centre cannot be infinite. Therefore, based on the arguments presented by al-Ghazālī, which implied that the universe could have been created larger or smaller than its known size, we now conclude that the philosophers should have abandoned, either their assumption of the eternity of the world, or their doctrine of a geocentric universe. It would be fascinating to see how this conclusion echoes in the modern understanding of the cosmos, a question which I leave for further research.

SCIENTIFIC ASSESSMENT

By the beginning of the 20th century, astronomers had started a programme of observations aimed at studying the motion of nearby galaxies. It was found

359 Ibid. 66.

that most of these galaxies, which are called "the local group", are descending away from us. Through patient observations made during the 1920s, it was established by Vesto Slipher and Edwin Hubble that the universe is in fact expanding. Hubble deduced that the further away a galaxy is from us, the faster it is descending.[360] Following this discovery, George Gamow and his collaborators developed a concept explaining the natural abundance of elements; that is the average percentage of each of the ninety-two natural elements found in the universe. This scenario was later called the "Big Bang Theory."

A continuously expanding universe was already a possibility suggested by the Theory of General Relativity. This theory was proposed by Albert Einstein in 1915 and, having been confirmed by many observations, it was adopted to be the standard theory of space, time, and gravity. The theory replaced Newton's law of gravity, which had served the astronomical calculation of the solar system for about 300 years. Almost all models of modern cosmology are based on this theory, according to which the universe is being driven to expansion by its own internal energy. Indeed, modern cosmology allows for an infinite universe as a possible solution to the Einstein field equations, although the universal model, which was proposed by Einstein himself, was static, finite, but unbound.

Einstein's static model was a sort of artefact designed by him after the modification of his field equations. Einstein was driven by the prevailing belief that the universe was finite and static, a belief that might be a relic of Aristotle's universe. The Einstein universe cannot expand, nor can it collapse, because once it starts to expand, it will do so forever and once it begins to shrink, it will go on shrinking towards a point. This critical behaviour makes Einstein's universe extremely unstable; like a pencil standing on its tip.

It is interesting to note that Ibn Rushd's conjecture concerning the ever-expanding universe finds echoes in Einstein's model. However, once the discoveries made by Hubble and others confirmed an expanding universe, Einstein's static universe became redundant. Other dynamic models, deduced using solutions of the original (unmodified) Einstein field equations, have been alternatively proposed. These provide us with three options:

1. A universe which expands forever with an ever-accelerating rate; this is called the "open universe."
2. A universe which expands forever but with a lower rate of acceleration, to reach an ultimate terminal speed at a later time; this is called the "flat universe."
3. The third model is a universe that expands until it reaches a maximum size in a finite duration of time and then begins to collapse, at the end

360 Edwin Hubble, *American Scientist* 43, no. 2 (1943).

> of which phase it returns to its original state; this is called the "closed universe."

It is this third model that may correspond with what the Qur'an points to in Surah Al-Anbya (21:104).

However, if the universe is expanding now, then this means that, in the immediate past, it must have been smaller in size. Therefore, one might ask, where is the universe expanding to? Is there a void beyond the universe into which it is expanding? Modern cosmology, which is based on the Theory of General Relativity, assumes that the universe is four-dimensional, three dimensions are for space and the fourth dimension is time, into which the universe is expanding. Thus, it appears that the universe has no 'outside', and if we have to talk about the universal volume in space, then we have to accept the fact that we can only see the surface of the universe from within. The cosmological model for the universe was set forth by the Theory of General Relativity, which states that the volume of three-dimensional space we see, is actually a three-dimensional surface embodied in a four-dimensional spacetime. Hence, time is the axis along which space is expanding. For this reason, cosmological expansion is understood as being the growth of space in between large cosmological structures. This allows us to view the situation as an analogy to the expansion of a two-dimensional balloon, where we see dots on its surface becoming separated by greater distances as the balloon is inflated.

It might be astonishing to learn that al-Ghazālī had come to the conclusion that the universe had no outside. He expressed his understanding thus, "If you mean by it anything else, then the world has no 'before,' just as when one means by 'outside the world' [something] other than its surface, then one would say, there is no exterior to the world."[361] This sentence was written in the context of stating that, although the world has a beginning, there is no instant before that beginning, thereby stressing the notion that space and time came into existence at the creation of the world but not before. Furthermore, al-Ghazālī treated space and time on an equal footing:

> It is thus established that beyond the world there is neither void nor filled space, even though the estimation does not acquiesce to accepting [this]. Similarly, it will be said that just as spatial extension is a concomitant of body, temporal extension is a concomitant of motion... There is no difference between temporal extension that in relation [to us] divides verbally into 'before' and 'after', and spatial extension that in relation [to us] divides into 'above' and 'below.' If then, it is legitimate to affirm an 'above' that has no above, it is legitimate to affirm a 'before' that has no real before, except an estimative imaginary [one] as with the 'above.'[362]

361 Al-Ghazālī, *Incoherence*, 35.

362 Ibid. 33.

This is surely an advanced conceptual understanding that is in agreement with the current understanding of modern cosmology and the Theory of General Relativity, which stipulates that space and time were born along with the universe and that there is no space beyond the surface of the world (which in general relativity terms is a three-dimensional surface).

THE DECAY OF THE SUN

The Sun is the brightest object in the sky. With its influence on terrestrial life on Earth, it has attracted the attention of man since the very early times of his existence. Some cultures worshipped the Sun and, on many occasions, it was taken to symbolise power and life. Here I will use the words corruption, decay and diminish interchangeably to give the same meaning.

Al-Ghazālī's Refutation of Galen's Thesis

According to al-Ghazālī, the Greek philosopher, Galen, proposed that the Sun was an eternal heavenly body that should not corrupt or decay. The notion that heavenly bodies were incorruptible was a basic doctrine of Aristotle and his followers.[363] The Sun, moon, planets, and stars were alleged to be formed of a fifth element called 'ether.' It was the sub-lunar world only—air and the Earth—which was believed to be corruptible.

In the second discussion of *Tahāfut al-falāsifah*, al-Ghazālī tried to refute the proposition put forward by Greek philosophers that the world, space, and time are eternal. Post-eternity of the world was the main issue in this discussion and, for this reason, he considered the example of the Sun's fate. The argument put forward by the philosophers (which al-Ghazālī attributes to Galen) said that: should the Sun diminish, it would suffer from withering, something which had not been observed during the many years of studying the Sun. Al-Ghazālī tried to refute this implicit pre-condition on the corruption of the Sun by suggesting that such a pre-condition was unnecessary, "But we do not concede that a thing is corrupted only by way of withering. Rather, withering is but one way of [a thing's] corruption."[364]

Furthermore, al-Ghazālī maintained that, even if the argument for withering was conceded, how would one know about withering except through astronomical observations? But, since astronomical observations were not so reliable, it would be impossible to detect a small diminishing in the size of the Sun. Al-Ghazālī explained that as the Sun was a very large object, a loss of a

363 I could not find a clear citation for Galen regarding this, but surely the general idea is known to be part of the Greek philosophical doctrine. See, for example, S. Marc Cohen, Patricia Curd, Charles David, and Chanel Reeve, eds., *Readings in Ancient Greek Philosophy: From Thales to Aristotle*, 3rd ed. (Indianapolis, IN: Hackett, 2005).
364 Al-Ghazālī, *Incoherence*, 49.

small part of it might go unnoticed, "Should the Sun, which is said to be a hundred and seventy times larger than the Earth, or close to this, be diminished by the size of mountains, for example, this would not be apparent to the senses...The senses, however, would have been unable to apprehend this because estimating [such an amount] is known in the science of optics only by approximation."[365] He then proceeded to make an analogy of the dissimilation of a ruby, whereby, it loses a very small amount of its mass over a long period of time, "This is similar to the case of rubies and gold that, according to [the philosophers], are composed of elements and are subject to corruption. If then a ruby is placed [somewhere] for a hundred years, what diminished of it would be imperceptible. Perhaps the ratio of what diminishes from the Sun during the period of the history of astronomical observations is the same as what diminishes of the ruby in a hundred years, this being something imperceptible."[366]

So, as we see here, al-Ghazālī not only believed in a corruptible Sun but conjectured that the Sun might actually be diminishing at a rate that would go unnoticed by the optical techniques available at his time, even by observations extending over a large period of time. This is what our current knowledge would certainly endorse.

Ibn Rushd Defending Galen's View

Ibn Rushd tried to defend Galen's view, claiming that, "Galen's statement is only of dialectical value."[367] He argued that if the heavens were to suffer such a major change as celestial objects becoming corrupt, then such a corruption would produce a sixth element. He says, "Should heaven, however, lose its form and receive another, there would exist a sixth element opposed to all the others, being neither heaven, nor earth, nor water, nor air, nor fire. And all this is impossible." This, he said, because the fifth heavenly element (ether) is supposed to be incorruptible, according to Greek philosophy. So, if it were to suffer corruption, then the element of which it is composed would have to change. As no such element had been identified in the composition of the world, thus for him such an element did not exist.

Ibn Rushd questioned, further, the possibility of the Sun decaying, wondering whether the secondary effects produced by the decay would affect the sub-lunar world, "If the Sun had decayed and the parts of it which had disintegrated during the period of its observation were imperceptible because of the size of its body, still the effect of its decay on bodies in the sub-lunar world

365 Ibid.

366 Ibid., 49.

367 Statements of Ibn Rushd in response to al-Ghazālī's arguments are taken from his book: *The Incoherence of the Incoherence*, translated by Simon Van Den Berg, EJW Gibb Memorial Trust, Cambridge University Press, 1987, 75-76.

would be perceptible in a definite degree." This was a reasonable expectation, since a decaying object would certainly produce some output that could be traced in the world through their secondary effects. The reason why such secondary effects are expected to happen is because, "For everything that decays does so only through the corruption and disintegration of its parts, and those parts which disconnect themselves from the decaying mass must necessarily remain in the world in their totality or change into other parts, and in either case, an appreciable change must occur in the world, either in the number or in the character of its parts."

In this statement, Ibn Rushd is expressing the law of the conservation of matter,[368] a notion which is so clear and bold here that it makes one admire his genius. However, for him, such an effect had not been observed and this, therefore, supported the proposition that the Sun does not corrupt. Furthermore, Ibn Rushd concluded his response to al-Ghazālī by resorting to a metaphysical argument, "To imagine, therefore, a dissipation of the heavenly bodies is to admit disarrangement in the divine order which, according to the philosopher, prevails in this world." In fact, this was not much of an argument since we cannot see how the divine order would become disarranged, unless we believe that the metaphysical order requires the heavens to be immune to corruption or change. This was what Ibn Rushd believed; that literally any change would cause disarrangement and an alteration of the divine order.

Scientific Assessment

Modern astrophysics has shown that the Sun, and indeed all other stars in the universe, generate a tremendous amount of energy through the process of nuclear fusion. This happens when four protons (a hydrogen nucleus) fuse at a high temperature and pressure, producing one helium nucleus. Consequently, a large amount of energy is released from the core of the Sun in the form of heat, light, and other radiation. According to the law of mass-energy equivalence, which was established by Albert Einstein, the amount of energy radiated by the Sun every second, in the form of heat, light, and other radiation, is equivalent to 4.2 million tons of mass. However, this radiation is only a small portion of the Sun's immense mass. At this rate, the Sun loses only about 0.001% of its mass every 150 million years. The Sun is believed to have a sufficient amount of hydrogen to sustain its energy production for the next five billion years or so, by which time the useful percentage of hydrogen will have been exhausted. Following this, the Sun would undergo a series of changes that will involve the fusion of a helium nucleus with carbon and oxygen molecules.

368 Although it is not so for the conservation of energy, as the concept of energy was unknown at the time.

A huge amount of energy would be released during this explosive fusion, causing the Sun to expand tremendously, thereby increasing its size by 100 and changing it into a "red giant." This late phase constitutes only a relatively short duration of the Sun's life, following which it will end up collapsing into its final fate as a tiny "white dwarf", which would be only just visible from Earth. This happens as the red giant cools, and the generation of heat and pressure ceases. Consequently, the Sun cannot sustain itself against the gravitational pull of its parts, causing it to collapse in a colossal event to become a white dwarf, with a size smaller than that of Earth, and glowing with only a faint light.

All stars that have approximately the same mass as the Sun will undergo a similar fate. Other, more massive stars develop into neutron stars; objects composed mainly of a neutron core and a radius of only about 10km. Stars which are more than 3.4 solar masses, will continue the course of their collapse and become black holes; entities with such a strong gravity that even light cannot escape them. Accordingly, it is reasonable to conclude that the view of al-Ghazālī is more realistic than the one expressed by Ibn Rushd, despite the very interesting objections that the latter raised against al-Ghazālī's arguments.

AL-GHAZĀLĪ'S POSITION ON SCIENCE AND RELIGION

On many occasions, we read that al-Ghazālī was against science and scientific thinking. Recently, a well-known physicist, Steven Weinberg,[369] claimed that al-Ghazālī was one of the main reasons for the decline in science and scientific thinking in the Islamic world. Although, Jamil Ragep has already responded to these accusations,[370] I will present excerpts from al-Ghazālī's Introduction in *Tahāfut al-falāsifah*, which show that he actually stood by the exact sciences and proper scientific thinking, while opposing philosophers and the atheistic view of the world. There are several other places where al-Ghazālī expressed his respect for the exact sciences, but these can be reported on another occasion.

Al-Ghazālī introduced his book, *Tahāfut al-falāsifah*, with a prologue in three parts. In the first part, he wrote about the main addressees of his book, namely Aristotle and Plato:

> Let us then restrict ourselves to showing the contradictions in the opinion of their leader, who is the philosopher *par excellence* and 'the first teacher.' For he has, as they claim, organised and refined their sciences, removed the redundant in their views and selected what is closest to the principles of their capricious beliefs, namely, Aristotle.[371]

369 Steven Weinberg, Times Literary Supplement, 17 January 2007.
370 Ragep, F. Jamil, When did Islamic Science die. *Viewpoint*, Newsletter of the British Society for the History of Science, No. 85, Feb. 2008.
371 Al-Ghazālī, *Incoherence*, 6.

In the second part, al-Ghazālī differentiated between those subjects of philosophy that he was targeting and those he was not:

> One into the refutation of which we shall not plunge, since this would serve no purpose. Whoever thinks that, to engage in a disputation for refuting such a theory is a religious duty, harms religion, and weakens it. For these matters rest on demonstrations; geometrical and arithmetical, that leaves no room for doubt.[372]

At this point, al-Ghazālī went even further by discussing some of the dogmatic suspicions among Muslims about scientific achievements and the possible claims that they might be in contradiction with the stipulations of the Qur'an and the teachings of the Prophet:

> When one studies these demonstrations and ascertains their proofs, deriving, thereby, information about the time of the two eclipses [and] their extent and duration, is told that this is contrary to religion, [such an individual] will not suspect this [science], only religion. The harm inflicted on religion by those who defend it, not by its proper way, is greater than [the harm caused by] those who attack it in the way proper to it.[373]

Al-Ghazālī questions the claims that the views on this matter conflict with religious teachings, maintaining, rather, that an in-depth understanding of these teachings does not contradict scientific methodologies and results:

> If it is reported that God's messenger (God's prayers and peace be upon him) said, 'The sun and moon are two of God's signs that are eclipsed neither for the death nor the life of anyone; should you witness such [events], then hasten to the remembrance of God and prayer.' How, then, does this agree with what [the philosophers] state? We say: there is nothing in this that contradicts what they have stated since there is nothing in it except the denial of the occurrence of the eclipse for the death or life of anyone, and the command to pray when it occurs. Why should it be so remote for religious law, which commands prayer at noon and sunset, to command as recommendable prayer at the occurrence of an eclipse?.[374]

Clearly, the above examples, which I have presented here at length, reflect al-Ghazālī's positive impression of exact scientific methods and calculations that are not, and should never be, in conflict with the proper understanding of religious teachings. I hope this will partly repudiate the infamous claims spread in the West that al-Ghazālī was against science and that he was one important reason for the decline of scientific pursuit in the Islamic world.

---

372 Ibid.

373 Ibid. 6.

374 Ibid.

REFLECTIONS

In reflecting on the two main problems presented in this chapter, one can say that al-Ghazālī presented arguments which may be summarised by saying that there is no reason why it should not be possible for the universe to have been created smaller or larger in size. It is true that al-Ghazālī brought this question under the auspices of God's ability; however, his main intention was not to question God's ability, but to question the status of the space beyond the world, if any. He actually intended to confuse the philosophers on this question, as they claimed that their approach satisfied the omnipotence of God. For this reason, we find that Ibn Rushd confirmed the philosophers' belief, from the perspective of God's ability to do whatever he wishes, within the canonical framework of creation. Al-Ghazālī, it seems, was aware of such an attitude, and for this reason he took the question further to puzzle the philosophers on the question of the designations of the *after* and the *before*. Obviously, al-Ghazālī had no knowledge about the expansion of the universe, nor did he conjecture about such an expansion. For this reason, the question that followed in connection with this argument was related to the recognition of a temporal succession of events which marked a beginning for time; a point with which al-Ghazālī used to refute the eternity of the world, as claimed by the philosophers. As far as I know, this problem, along with other issues raised by al-Ghazālī, have not yet been studied, even though, as is shown in the related arguments and the concluded views here, they have a sound value in modern cosmology (although, it may not have been al-Ghazālī's intention to claim such a target).

The second problem was the question concerning the post-eternity of the world, for which al-Ghazālī took the example of the post-eternity of the Sun. He posed the question of whether the Sun suffers any corruption over time, a point which was pivotal in Greek philosophy. This question was directly related to the classification of the world into corruptible and incorruptible parts. Aristotle had classified the heavenly bodies as being incorruptible and, therefore, raising this point was of great important to al-Ghazālī, in order to demolish this classification. In fact, some Muslim theologians and well-known Mutakallimūn maintained that the heavenly bodies were of a different chemical composition to that of the Earth.

Al-Bāqillānī, the grand mentor of al-Ghazālī, clearly rejected the notion of ethereal celestial bodies, "As for those saying that celestial bodies are of a fifth nature, not fire nor earth, air nor water, [I would say that] this is flawed and has no proof."[375] Moreover, we see that al-Bāqillānī, who rejected the notion of

375 Al-Bāqillānī, *Tamhīd*, 64.

the four basic elements and their intrinsic natures, also rejected astrology on a rational basis and denied any effect of the celestial bodies on the Earth and its constituents. We find him writing:

> If someone were to say, 'why do you deny that the Maker of this world and His performer, ruler… could have been the seven spheres that are the Sun, Moon, Saturn, Mars, Jupiter, Venus and Mercury?', we would say: 'we deny that because we know that these stars are created and they are following the course of other objects in the world since it has similar constraints of limits, finiteness, composition, motion, rest and change from one state into another which applies to all other bodies of the world. Thus, if it were to be eternal, all other objects should be eternal too.[376]

In other places, al-Bāqillānī counters claims for astrological bodies emerging as generative influences on the basis that all celestial objects have the same qualities, "If it would be acceptable for these effects to be generated, then the Sun should generate the same effects as those generated by the Moon, and solid rocks should generate the same effects as generated by those celestial spheres, since they are all of the same quality."[377] Here again, we find that the Mutakallimūn presented an advanced view of the world, making the point that it is one and the same with respect to its basic constituents and the laws that are in action. The reason that the Mutakallimūn refused to attribute actions to inanimate matter is because such actions can only be generated by the presence of a will and reason. They denied that inanimate matter could have any kind of will or reason.

Ibn Rushd discussed the arguments put forward by al-Ghazālī regarding the size of the universe and the corruption of the Sun, and tried to show that these arguments were faulty. Obviously, Ibn Rushd relied completely on Aristotelian views and syllogism. He tried in vain, as far as I can see, to convince the readers that the arguments of al-Ghazālī were not valid, since his thinking went outside the existing framework. This might be true and it might have convinced a limited circle of thinkers, but not those outside it, and surely not the contemporary scientists and philosophers. The views presented by Ibn Rushd concerning these two problems would have been acceptable within the context of pre-Galilean physics, but certainly not in astrophysics and modern cosmology.

376 Ibid. 66.
377 Ibid. 75.

Chapter Twelve

The Anthropic Principle and the Multiverse

In the modern scientific view, we understand that the vast space containing the heavenly bodies is only the volume that covers what we now call the universe. Using optical and radio telescopes, we are able to explore the universe with a high degree of accuracy, covering distances which extend beyond billions of light years. Astronomers are able to analyse the light coming from luminous objects located at these distances and learn, in great detail, about their chemical composition. Currently, great efforts are being made to improve upon these techniques and the information received has introduced new facts which confirm the existence of an interconnected creation and make for a balanced evaluation of every single part of the universe.

The question of mankind's position in the universe has gained much importance in the last few decades, especially following the discovery of the peculiar adjustments required by the laws of nature in order to make the existence of humankind on Earth explainable by the laws of physics. The possibility of developed alien life systems existing in another part of the universe appears to be remote. Despite the fact that we are now sure of the existence of other planets belonging to extra-solar systems within our galaxy and other galaxies, we are uncertain whether they are capable of harbouring any life-forms.

Scientific speculation offers high expectations, but the extremely difficult conditions required for developing complicated life systems like ours, makes the probability of finding such a planet very low.[378] Nevertheless, we are now sure of one fact: the construction and development of the whole universe is firmly connected with the possibility of developing life somewhere in this universe. A profound mediation, both in experimental results and theoretical explanations has been set for the possibility of these discoveries, showing that all the parts are connected together. This reflects the fact that there are some strange similarities and precise relationships between the smallest and largest in this universe. This indicates a system that has been formed by a sort of cosmic conspiracy in order to make the existence of humans in such a universe possible. To gain an idea of what I mean here, I shall review some of the relationships between the basic constituents of the universe and the major forces at play.

THE NEUTRINO

This is a neutral particle with no electric charge and is one of the basic constituents of the atomic nucleus. It was assumed to exist when studies conducted on the products of elementary particle interactions showed that the equations could not be balanced without allowing for its presence. Its properties were discovered as a result of particle accelerator experiments, whereby subatomic particles collided at high speeds, causing them to break up such that their internal structures and contents could be investigated.

Elementary particles appear as traces on detection devices, such as the Cloud Chamber, where steam molecules condense along the path of the charged particle. Uncharged particles do not show such traces, making it difficult to discover them directly. This is why the researchers were unable to prove the existence of the neutrino. However, researchers were able to show that these particles, which were formed immediately after the Big Bang, make up a large proportion of particles in the universe. In fact, it is estimated that there are about 10^9 (i.e. a thousand million) neutrinos for each proton and electron in the universe.

It is believed that these particles' interaction with other materials is very weak. For example, the Earth is almost transparent to them, allowing them to pass through it easily. However, due to their vast numbers, the whole structure of the universe became very sensitive to their presence. It was generally believed that neutrinos were massless particles (i.e. energy waves), so that physicists assumed that they moved as fast as light. Experiments, conducted at the

378 Meadows, V. S., *Planetary Environmental Signatures for Habitability and Life*, in Mason, J. (Editor), Exoplanets: detection, Formation, Properties, Habitability, Springer, (2008).

beginning of the 1980s, showed that neutrinos might have some rest-mass, which is estimated at a few parts in 10^{-37} kilograms (i.e. 5 parts in 10 million parts of the mass of the electron). As we know that the electron is the smallest mass particle, we can, then, appreciate just how small the mass of a neutrino is. The high-density of these particles (about one billion particles in each cubic meter) means that their total mass is more than the mass of all the stars in the universe.

In order to understand how the mass of such a tiny particle affects the structure of the early universe, we can say that if the mass of these particles was a little greater, it could make a substantial change in the speed of the universe's expansion, or completely halt it. It could even have caused the universe to contract before the existence of man on Earth became possible. On the other hand, if the mass of the neutrino was much heavier, they would have gathered at the centre of the galaxy because their escape velocity would be greater than their normal speed. This would have resulted in their forming a kind of heavy fog, which would obstruct the rotation of the galaxy, thereby halting its universal movement.[379]

The calculations made by a team of theoretical physicists at the University of Texas in Austin, in the United States, confirmed that any simple change in the neutrino mass would cause a serious cracking in the structure of the galaxy. In addition to all this, the neutrino plays a very important role in balancing the energies and spin of nuclear particles during their interactions, thus playing a highly sensitive role in defining the ratio of these particles in the structure of the universe from its early beginnings. If the interaction's power factor was a little larger or smaller than what it is, then there would not have been enough hydrogen in the universe.

HYDROGEN

Hydrogen plays a significant role in the chemistry of the universe. Without it there would not be any organic materials or water on Earth, or anywhere else in the universe. Nor would there be enough fuel for the burning stars, such as the Sun. Consequently, there would be no life in the universe.

The normal hydrogen atom (the protium) consists of one proton and one electron, but there are two other hydrogen isotopes: deuterium, with one proton, one electron and one neutron in its nucleus, and tritium, with one proton, one electron and two neutrons in its nucleus. It was known that the deuteron (a deuterium nucleus) is an important component of the nuclear fusion cycle which occurs inside the Sun. This fusion reaction produces helium and an excess of the energy, which makes the Sun shine.

379 Davies, P.C.W. *The Accidental Universe*, 61.

Hydrogen isotopes have the same chemical properties as normal hydrogen, but they are different in their physical properties. The mass of the neutron is larger than the mass of the proton by about 1/1000 of the proton's mass. This is why the free neutron can decompose into a proton, an electron and antineutrino. If the difference between the mass of the neutron and the mass of the proton was any smaller (e.g., one-third of its value), then the free neutrons would be unable to decompose into protons as they would have an insufficient mass to generate the necessary electron with which to balance the charge. This would lead to a fundamental change in the quality of nuclear interactions in the universe. If the neutron mass was 99.8% of its current value, a free proton would decay into a neutron and a positron (positive electrons) and, consequently, there would not be any hydrogen atoms in the universe. This would mean, again, that there could be no possibility for life as we know it.

NUCLEAR FORCES

Nuclear forces play a significant role in the structure of the nuclei of chemical elements and, indeed, of the whole universe. The helium nucleus contains two protons and one neutron (helium-3) or two neutron and two protons (helium-4), which is the dominant helium isotope in the universe. Since similar charges repel each other, the electrical force between two protons in an atom would push them away from each other with a tremendous force which amounts to more than ten million billion tons. However, nuclear forces bind the protons and neutrons found within the minute space inside the atomic nucleus because they work at very short distances. As such, nuclear forces are much stronger than electrical forces; strong enough, in fact, to easily overcome the forces of the electrical repulsion. Because of their short range, nuclear forces act on the neighbouring particles while, due to their long range, electrical forces act on all charged particles.[380]

This means that any proton in the nucleus is bound to its neighbour by nuclear forces, which pull it to the nearest distance they can, while at the same time, it is being pushed away by the farthest protons with their electrical repulsion. In light nuclei, which contain a small number of protons, the repulsion powers have a negligible effect; however, in the large nuclei that contain a large number of protons, electrical repulsion is much more effective. This causes heavy nuclei to disintegrate spontaneously, producing radioactivity. In this way, the nucleus gets rid of the heavy stuff inside it to become lighter and more stable.

380 Povh, B.; Rith, K.; Scholz, C.; Zetsche, F. (2002). *Particles and Nuclei: An Introduction to the Physical Concepts*. Berlin: Springer-Verlag, 73.

If the nuclear forces were a little weaker, there would be only a small amount of stable chemical elements. If the value of any nuclear coupling constant is half of its current value, then the nucleus of, say iron or even carbon would not be stable for very long. This would lead to the loss of one of the most important elements for life, since carbon is an essential component of DNA (deoxyribonucleic acid), the molecule which contains the genetic code (as chromosomes) i.e. the propagation of life would be impossible.

What would happen if nuclear forces, which hold the nucleus together, were to be cancelled? The answer is that the universe, as we know it, would vanish, and instead, be turned into flying protons and neutrons. In addition, there would be a very significant amount of hydrogen and no helium, i.e. the universe would be a silent, static and meaningless place, with no possibility of producing life.

CARBON AND OXYGEN

Since carbon and oxygen are essential elements in the cells of living organisms, the British astrophysicist Fred Hoyle gave great attention to understanding how these elements formed stars. Carbon is generated from combining three helium atoms, but this harmonious combination would be naturally rare without the necessary balanced arrangements. This is because the fusion of two helium nuclei produces an unstable beryllium nucleus. The possibility of a third helium nucleus fusing to make a carbon atom before the unstable beryllium nucleus decayed depends on the energy by which the helium nucleus hits the temporary beryllium nucleus. This is because of the nuclear resonance requirement. When the quantum wave function of the helium nucleus becomes compatible with the internal vibration of the unstable beryllium nucleus, the cross section to combine with the third helium nucleus will sharply rise.

Incidentally, the thermal energy of the nuclear constituents of a typical star is located exactly at the point of resonance in the carbon atom.[381] This wise adjustment of natural properties results in a low probability resonance such that sufficient carbon atoms are produced inside the stars. Without this, the abundance of carbon in the universe would be very low and no carbon-based life would be formed.

Carbon can be converted into oxygen during nuclear reactions, but because the nuclear resonance in an oxygen nucleus is well below the thermal level of the constituents, most of the carbon is saved from this fate. Nuclear reactions are complicated, however, and the location of nuclear resonances depends on the primary strength of fundamental forces in the universe, especially

381 Salaris, Maurizio; Cassisi, Santi (2005), *Evolution of stars and stellar populations*, John Wiley and Sons, 119–121,

nuclear and electromagnetic forces. If the strength of these forces is different, the precise arrangements in the resonance of carbon and oxygen would not happen and the possibility of life on Earth would be negligible.[382]

Does this mean that life depends only on carbon? Absolutely not. Scientists proposed other alternatives for life depending on which other elements were utilised, e.g. nitrogen, although, of course, this would result in a new type of creation. This means that, creating life in another form or via a different developmental path, would require a huge change in the universe, which no one can do but the One who performed the first creation.

GRAVITATIONAL AND ELECTRICAL FORCES

It is noticed that gravitational force has universal domination as it has a tremendously long-range effect; infinite, in fact. This force controls the universe and the movement of far galaxies such that even light—which is so fast that it can circle the Earth seven times in one second—needs billions of years to reach us from them. The electromagnetic forces which dominate the microscopic world of atoms and molecules are much stronger than the force of gravity. To demonstrate the difference between gravitational and electrical forces in the nuclear realm, we must mention that the electrical force between the electron and the proton in a hydrogen atom is 10^{40} times stronger than the gravitational force between them!

The strength of these two forces depends on the fine structure constant of each, where the strength of certain forces reflects the value of this constant (commonly denoted α). The value of the constant, with respect to gravitational force, is 5.9×10^{-39}, while its value, with regards to electromagnetic power, is 7.3×10^{-3}. In fact, the determination of these values has a great influence on the organisation of the universe, as well as in the creation of typical stars, and their stabilisation as they produce light and heat.[383] A good example of this is the Sun which, without such precisely determined values for the strength of the forces would not be viable in its current form. The structure of a star depends on its ability to spill out the heat from its core onto its surface through radiation. In massive stars, the so-called blue giants, radiation energy becomes the dominant energy, with thermal energy dissipating from them through radiation.

If the mass of the star is smaller, this way of moving energy would not be feasible because radiation cannot travel fast enough as to maintain the surface of the star with sufficient heat. This is an important point; if there are an inadequate number of ions on the surface of the star, it will become unstable, and its heat removed by convection.

---

382 Ostlie, D.A. & Carroll, B.W. (2007). *An Introduction to Modern Stellar Astrophysics*. Addison Wesley, San Francisco.

383 Davies, *The Accidental Universe*, 48.

The convectional excitation of heat will be complemented by the flow of radiation energy and will prevent the temperature from being reduced substantially below the ionisation temperature. This is why those stars which use thermal convection methods are smaller and cooler than blue giants. Such stars are called red giants. The Sun, categorised as a yellow dwarf, and some other stable stars fall between the bands of these two-star classifications.

THE ACCIDENT AND THE PURPOSE

Brandon Carter, a famous astronomer, proved that if gravitational forces were a little weaker, electromagnetic forces a little stronger or electron mass a little smaller in proportion to proton mass, then all stars would be red dwarfs. Accordingly, a change in the opposite direction would turn all the stars into blue giants. Consequently, a world with slightly lower gravity might not contain planets. In both cases, if gravity was a little more or a little less, then our universe would be totally different from the way it is now.

The facts introduced above have attracted the attention of scientists working in cosmology, atomic physics and astronomy to the possibility that an intelligent power may stand behind the creation of the universe. Some scientists found it naïve to adopt the idea that the birth of the universe could be accidental. The probability of an accidental universe being created is extremely low, and to say that many accidental chances have occurred coincidentally to make this universe possible is unreasonable. Among the numerous options, the possibility of the emergence of life, particularly intelligent life, is significantly low. How could all these constants and factors have been arranged with such values as to lead to the emergence of intelligent life and a level of consciousness that we enjoy? What sort of conspiracy is behind the scenes?

In the last two decades, the fact of the finely tuned universe was exposed when theoretical research on cosmic microwave background radiation revealed the very accurate tuning of the initial conditions for creating this universe. The expansion rate of the universe depends on the value of the cosmological constant and this, in turn, depends on the balance between the respective contributions of the quantum vacuum and the bare cosmological constant. The author has worked on these topics and is aware that the balance between the vacuum energy and the bare cosmological constant can be up to 10^{-50}.

THE ANTHROPIC PRINCIPLE

From the above, and many other observations, scientists realised that there was a strong relationship between every single particle in the universe and the existence of humans on an Earth that moves around one specific star out of millions of stars in the whole universe. Yes, there might be other planets which harbour life, but they have not yet been discovered.

The fact that we are here in this universe, in spite of the odds stacked against us, induced scientists to attempt to find an explanation for those coincidences which make our existence possible. I admit that this is one of the hardest questions that one may face, and most geniuses in the scientific world have expressed their astonishment as to how our mental ability can be in such harmony with the universe that we can comprehend it perfectly. Einstein is quoted to have said, "The most incomprehensible thing about the universe is that it is comprehensible."[384]

This strong relationship between how the universe began and our highly developed and complex presence introduced a new concept into the mix; known as the anthropic principle. This notion emerged as an explanation for the observed constancy of nature and the harmony of the laws of nature with the emergence of intelligent life. Scientists have expressed two main views about this principle.

The Weak Anthropic Principle

In one version of it, this principle says that our existence is just a happy, rare chance by which sporadic natural incidences have combined over a very long period of time to make our existence possible. This is the weak anthropic principle, defined by Brandon Carter as follows, "what we can expect to observe must be restricted by the conditions necessary for our presence as observers."[385] Followers of this explanation admit that it was a rare opportunity to have intelligent life developing on Earth, but they consider it some sort of factual reality that we are here. Which translated means: that if there could not be such a "happy chance", we would not be here to talk about it. This what we understand from the words of Paul Davies when he says that Hoyle's example of the formation of carbon in stars through the right balance of resonance and energy "does not explain the coincidence of nuclear energies, but merely comments on the extreme fortune of the circumstances: had it not been so, we should not have been here to discuss it."[386] But here Davies is overlooking the fact that Hoyle's illustration involves bringing in the two requirements which are not inter-dependent (that is, resonance and the star's temperature), in order to achieve the goal of producing enough carbon necessary for our existence.

Other scientists, such as Steven Weinberg, have tried to devaluate the importance of Hoyle's remark by overestimating the probability of achieving the required resonances by claiming that it is not as rare as it might be thought.

384 Einstein, A. (1936), *Physics and Reality, in Ideas and Opinions*, trans. Sonja Bargmann (New York: Bonanza, 1954), 292.

385 Brandon Carter, (1974) in *Confrontation of Cosmological Theories with Observation* (ed. M.S. Longair, Reidel, Dordrecht).

386 Davies, *The Accidental Universe*, 119.

Nevertheless, Weinberg finds no way of escaping the fact of fine tuning on other occasions, such as the precise adjustment of the value of the cosmological constant.[387]

The Strong Anthropic Principle

This version of the principle was proposed as evidence of a deliberate organisation and design intended to make our existence possible. Brandon Carter called it the strong anthropic principle and described it thus, "The universe must be such as to admit the creation of observers within it at some stage."[388] This statement implies that the universe was tailor-made for habitation by creatures like us through the interplay of the laws of nature and the initial conditions, which arranged themselves in such a manner as to assure the presence of living organisms. Davies correctly recognises that, "the strong anthropic principle is akin to the traditional religious explanation of the world which says: that God made the world for mankind to inhabit."[389]

Many scientists showed support for the notion of the strong anthropic principle, among them Josef Silk, John Wheeler and John Barrow. Josef Silk remarked that, "Gravitational instability and fragmentation must lead from giant clusters to galaxies to stars, and ultimately to planets and an environment suitable for the development of life."[390] Wheeler points to the fact that such a large universe is necessary for our existence and, unless the radius of the cosmic horizon is more than 10^9 light years, the universe will collapse. John Barrow and Frank Tipler jointly authored a large volume in which they supported the strong anthropic principle, pointing to the coincidences of the physical values and the action of the laws, saying that, "The possibility of our own existence seems to hinge precariously upon these coincidences. These relationships, and many other peculiar aspects of the universe's make-up, appear to be necessary to allow the evolution of carbon-based organisms like ourselves."[391]

The primary remarkable fact is that our existence is conditional upon a delicate physical relationship and the values of certain fundamental physical constants, which occur through infrequent chance. There are two possible explanations for why such a rare chance would take place: one is that the universe has been around for an infinite amount of time, which allowed for the

---

387 Weinberg, Steven. "The cosmological constant problems." In *Sources and Detection of Dark Matter and Dark Energy in the Universe*, pp. 18-26. Springer, Berlin, Heidelberg, 2001.

388 Carter, B. (1974). "Large Number Coincidences and the Anthropic Principle in Cosmology." IAU Symposium 63: Confrontation of Cosmological Theories with observational Data. Dordrecht: Reidel. pp. 291–298.

389 Davies, *The Accidental Universe*, 121.

390 Josef Silk, Cosmogony and the magnitude of the dimensionless gravitational coupling constant, Nature, 265, 710, 1977.

391 Barrow and Tipler, *Anthropic Cosmological Principle*, xi.

emergence of a high level of cosmic organisation. This might be thought of as the result of a stupendously unusual statistical fluctuation from the far more probable condition of featureless disorder. This explanation was conjured up by Ludwig Boltzmann, who suggested that a cosmic-sized version of the fluctuations that produce a Brownian motion could create the optimum conditions required for those rare events necessary for our existence. This requires a universe whose age extends to a hundred billion, or even billions of billions, of years, which is out of the scope of present cosmological observations. The second option is to assume the existence of many universes (the multiverse proposal), out of which our universe happens to have enjoyed such rare conditions as to make our existence possible.

Davies has observed that, "Given the random cavorting of all atoms, wholesale cooperation of large numbers of atoms will conspire, after unbelievably long duration, to produce order spontaneously out of chaos."[392] The point to remark here is that the explanation given by the weak anthropic principle is no explanation since it does not offer any reason as to why such a happy chance would occur. Why should there be so much co-ordination so as to make our existence possible? On the other hand, the so-called strong anthropic principle is no real explanation either. It only states the fact that such fine tuning of many parameters and such delicate adjustments are purposely intended to have us present in this universe. But why? Why should there be observers at all? Wouldn't this indicate that the presence of an observer is necessary for some purpose? What then is the necessity for this observer?

EXPLAINING THE ANTHROPIC PRINCIPLE

Fine tuning is an observable fact. Its connection with the existence of a conscious, intelligent observer who can comprehend it is what the anthropic principle is all about. But there should be a reason for fine tuning since it indicates that some co-ordination took place in order to adjust the universal constants and the fundamental physical quantities so as to achieve those precise values which make our existence in this world possible. Why should there be such precise fine tuning? Perhaps it would be sensible to say that this fine tuning is a necessary pre-condition for managing the world with *blind* laws rather than smart angels. Should such fine tuning not be available, the blind laws of nature[393] would not have been able to develop the universe sufficiently to have us here. Precise definitions, strict ruling and detailed instructions are always

392 Davies, *The Accidental Universe*, 128.

393 By the 'laws of nature' I mean all the natural phenomena. These are different from the laws of physics, which are normally our explanation for the laws of nature, see Basil Altaie, *God Nature and the Cause*, KRM, 2016, chapter 2.

needed whenever you have a group of *idiots* at work through your system. In this case, the universe can have two options:

1. Have no freedom for the blind laws at work, thus making them work deterministically, (Richard Dawkins would be pleased with this option), or
2. allow these laws some freedom to become indeterministic.

With the first option, we end up with a world ruled by the deterministic laws of nature and by which the universe will need no driver/coordinator (i.e. no God). But this only allows for limited diversity of creation and is against what quantum mechanics tells us. The second option allows for the universe to be ruled by indeterministic laws that work with some freedom, as John Polkinghorne predicted. However, this option necessitates the presence of a driver/coordinator to choose from the different contingent possibilities and produce a widely diversified world. The presence of the intelligent observer would justify the need to verify and testify to the grand purpose.

This explanation shows that, despite the fact we have blind, innate laws at work, the watch maker, after all, is not blind. It is not the tools that are making the parts of the watch and collecting them, but it is the one who is using those blind tools, who is actually making the watch. Those blind tools can see through their user in a way that Dawkins couldn't recognise.

THE MULTIVERSE HYPOTHESIS

The notion of multiverse is a convenient concept that lends itself to explain anything that one may find peculiar. By its own construct, being dependent on infinite possibilities, this notion of multiverse allows us to choose whatever we like out of such an unlimited set. Steven Weinberg,[394] for example, uses it to explain the value of the cosmological constant, which he relates to anthropic issues. The Astronomer Royal, Martin Rees,[395] employs it to explain the whole set of anthropic coincidences, that is, to explain why our universe is a congenial home for life. Others, such as John Barrow,[396] use it to speculate about the values of physical constants in our universe. In short, the notion of multiverse, as presented in the arena of science today, makes everything possible.

The basic idea of having multiple universes seems completely natural; we have many planets, many stars, many galaxies and many galactic clusters, so why have only one universe?

394 Weinberg, S., "The Cosmological Constant Problems," see footnote (28).

395 Rees, M. J. 2001, *Just Six Numbers: The Deep Forces that Shape the Universe* (Basic Books). Rees, M.J. 2001 *Our Cosmic Habitat*, (Princeton University Press). M J Rees 2001, 'Concluding Perspective', astro-ph/0101268.

396 Barrow, J. D. and J.K. Web, Inconstant Constants, *Scientific American*, June 2005.

There are three origins for the notion of multiverse: the first is the Everett-Wheeler interpretation of quantum mechanics, the second is chaotic inflation, proposed by Andre Linde,[397] and the third is string theory.

The general theory of relativity uncovered the integral nature of space and time as being interwoven into one spacetime continuum. Accordingly, we should have one universe. However, inflation theory suggests that there could be multiple, causally disconnected regions of spacetime, each forming an independent spacetime manifold; therefore, a universe on its own. This is the origin of the multiverse notion, derived from the theory of inflation.

Everett proposed that quantum states split, or branch, upon measurement, as an explanation trying to resolve the problem of measurement in quantum mechanics. This proposal triggered the notion that each branch is a universe on its own, thus producing multiverse (many worlds) after a sequence of measurements.

String theory brings in a similar idea, suggesting that the time development of string in higher dimensions would produce superstrings that, when taken in two dimensions, will form branes (short for membranes), which express the character of a multiverse.[398]

It is important to define what we mean by multiverse. This question has been raised by George Ellis, the renowned cosmologist and a leading expert on Einstein's general relativity, who has written several articles on evaluating the notion of multiverse. Ellis says, "There is however a vagueness about the proposed nature of multiverses."[399] In a recent article published in *scientific American*, he analysed the proposal of multiverse and concluded that, "We are going to have to live with that uncertainty. Nothing is wrong with scientifically based philosophical speculation, which is what multiverse proposals are. But we should name it for what it is."[400]

Astronomers have realised that what they see with their most powerful telescopes is not the whole universe but only part of it. This can extend, at most, to 42 billion light years. This is our cosmic visual horizon. But there is no reason to believe that the universe stops at such boundaries. Beyond it could be many more domains, similar to what we can see. Each might have a different initial distribution of matter, but the same laws of physics would

397 Linde, A. D. 1983, Physics Letters 129B, 177; Linde, A. D. 1990, Particle Physics and Inflationary Cosmology, Harwood Academic Publishers, Chur, Switzerland.

398 Szabo, Richard J, *An introduction to string theory and D-brane dynamics*, Imperial College Press; 2 edition, (2011).

399 Ellis, George F.R., Ulrich Kirchner, William R. Stoeger, Multiverses and Physical Cosmology, Mon. Not. Roy. Astron. Soc., 347 (2004) 921-936.

400 George F.R. Ellis, Does the Multiverse Really Exist? *Scientific American*, August 2011, 39-43.

operate in all. Most cosmologists today accept this type of multiverse, which Max Tegmark[401] calls Level 1. Yet some physicists, like Alexander Vilenkin, go much further in painting a dramatic picture of an infinite set of universes with an infinite number of galaxies, an infinite number of planets and an infinite number of people with your name who are reading this article. Needless to say, that these infinite universes are but mere possible mathematical solutions endorsed by equations which set the formulation for the proposed systems. Since the dawn of mathematics and algebra, we have known that, in many cases, algebraic equations can have an infinite number of solutions. Physicists usually pick those solutions which represent realistic states and ignore the others. The fact that general relativity has enabled us to write an equation for the whole of spacetime has given some physicists the excuse to interpret those different solutions as representing different universes that may exist. But the bizarre thing is that they are providing these solutions without applying any constraints, since they have none to apply. Accordingly, they suggest completely different types of universes, with different physics, different histories, and maybe even different numbers of spatial dimensions. Most will be sterile, although some will be teeming with life. This is what is called a "Level 2" multiverse.

Similar claims have been made since antiquity by many cultures. What is new is the assertion that the multiverse is a scientific theory, with all its implications being mathematically rigorous and experimentally testable. Even so, George Ellis says, "I am skeptical about this claim. I do not believe the existence of those other universes has been proved—or ever could be." [402]

There are various proposals as to how such a proliferation of universes might arise and where they would all reside. They might be sitting in regions of space far beyond our own, as envisaged by the chaotic inflation model of Alan Guth, Andre Linde and others.[403] On the other hand, Paul Steinhardt and Neil Turok[404] suggested that such multiverses may exist at different epochs within the cyclic universe model of universe. David Deutsch[405] advocates that they might exist in the same space we do, but in a different branch of the quantum wave function. Conversely, Max Tegmark[406] suggested they might not have a location, being completely disconnected from our spacetime.

---

401 Tegmark, M. (2003). "Parallel universes." *Scientific American*, 288(5), 40-51.

402 George Ellis, Opposing the multiverse, Astronomy & Geophysics, Volume 49, Issue 2, 1 April 2008, 2.33–2.35.

403 Andrei Linde, "The Self-Reproducing Inflationary Universe," *Scientific American*, November 1994.

404 Gabriele Veneziano, The Myth of the Beginning of Time, *Scientific American*, May 2004.

405 David Deutsch and Michael Lockwood, "The Quantum Physics of Time Travel", *Scientific American*, March 1994.

406 Max Tegmark, "Parallel Universes," *Scientific American*, May 2003.

For a cosmologist, the basic problem with all multiverse proposals is the presence of a cosmic visual horizon. The horizon is the limit of how far away we can see, because electromagnetic signals, which have been travelling at the speed of light (which is finite) since the creation of the universe, have not had enough time to reach us. All the parallel universes lie outside our horizon and remain beyond our capacity to see, no matter how much our technology evolves. In fact, they are too far away to have had any influence on our universe whatsoever. That is why none of the claims made by multiverse enthusiasts can be directly substantiated.

Let us look at the landscape of the planets in our universe. Suppose we want to extrapolate on their properties and the laws that apply to their control and development. Do we see that each planet has its own laws? No. If we examine the landscapes of the billions of galaxies in our universe, do we find different laws for each of their formations and developments? No. Why then should we allow for different laws of physics to be at work within other worlds and landscapes within the multiverse? It seems, to me, that such an adventure is unnecessary.

Looking through the recent defences for the multiverse theory, one can see that some proponents of the notion have become more modest regarding their speculations. Max Tegmark remarks that no one from the proponents of the multiverse says that these parallel universes really exist.[407] Therefore, let us stick to the fact that these are mere mathematical solutions and rely on objectivity to resolve the question of whether they exist or not. Objectivity tells us that we should not scrutinise speculations too closely once we have set a basis which allows us to take everything from our imaginations as possible. The open landscape of the multiverse is such a danger as it may take science into a new venture that goes beyond the logic by which we comprehend reality. We have to admit that objectivity tells us that the multiverse of Level 1 can be accepted as the only possibility which we can comprehend. In no way can we deal with other levels, since our knowledge is not equipped with the necessary ingredients needed for such a venture.

If we are to tackle such questions, we should break away from what we know and try to investigate such possibilities systematically. For example, we may think of a world with antigravity or a world working with an inverse of the second law of thermodynamics. We can play with variances of the laws of statistics and quantum mechanics in order to see what kind of outcomes we obtain. In this case, the hypothesis of multiverse could be a productive research program.

---

407 Max Tegmark, "Are Parallel Universes Unscientific Nonsense? Insider Tips for criticizing the multiverse," *Scientific American* February, 2014.

From the Islamic point of view, there is nothing of contradiction between belief in Islam and the existence of multiverse of any type, since it is assumed that Allah, the Creator, is able to create whatever He wills. Therefore, there is no limit on the number of worlds he can create and no restrictions on their properties. In fact, such worlds, even with different laws of physics, are deemed necessary to acknowledge the afterlife; be it a hell or a heaven. In some literature and religious narrations there is a detailed description of such worlds. Thus, in one way or another, the notion of multiverse may help give a scientific touch to some metaphysical religious thoughts.

Although the idea of parallel universes is attractive, I find it is still immature and unfit to answer many questions. Hereafter, I would like to present the Islamic understanding for the fine tuning of the universe.

FINE TUNING AND 'TASKHĪR' IN ISLAM

In Islam, we encounter a principle which says that whatever we see in the universe was arranged for our service and to make the purpose of our presence lucid. This is called the principle of subservience (*taskhīr*), a topic which finds its application in *Jalīl al-kalām.*

The Qur'an gives accurate descriptions for the construction of the world and the introduction of an intelligent creature who can contemplate and comprehend the universe in order to learn about the laws governing it. We first read that the world was created with the utmost care and that the values of its constituents were chosen with great accuracy. In Surah Ar-Rad:

> *And everything with Him [Allah] has a measure. (13:08)*

And Surah Al-Furqan:

> *He has created everything, and has measured it exactly according to its due measurements. (25:02)*

And Surah Al-Qamar:

> *Surely, We have created everything according to a measure. (54:49)*

These verses, and many others, assert that the creation was performed to a precise measure. This allows us to assume that such a measure is necessary for achieving a certain target. Furthermore, it implies that creation was achieved according to certain rules that require delicate measure and care.

The Qur'an reveals that the Creator made everything on Earth and in the Heavens subservient to mankind. In Surah Al-Jathiyah:

> *And He has made subservient (subjected) to you whatsoever is in the heavens and whatsoever is in the earth, all, from Himself. Surely there are signs in this for a people who reflect. (45:13)*

And Surah Luqman:

> *See you not that Allah has made subservient (subjected) to you whatever is in the heavens and whatever is in the earth, and granted to you His favours complete outwardly and inwardly? And among men is he who disputes concerning Allah without knowledge or guidance or a Book giving light. (31:20)*

And Surah An-Nahl:

> *And He has made subservient for you the night and the day and the sun and the moon. And the stars are made subservient by His command. Surely there are signs in this for a people who understand. (16:12)*

Indeed, these are signs for people who understand. It is worth mentioning that, whenever the Qur'an mentions subjecting or subservience, it mostly comes in the context of the general creation of the Heavens and the Earth. Thus, the intended subjection of the heavens, stars, sun and moon is related to the issue of creation, in which the organisation of creation is to be at man's service. To say that the appearance of mankind at a particular age of the universe is just a "lucky moment", is a superficial understanding of the anthropic principle. In fact, there are many other verses in the Qur'an which tell of subjecting the rivers, the winds, and water falling down to the ground, for the benefit of mankind. This is part of the subservience (*taskhīr*) extended for the life of humans. But why should humans be so important?

The answer lies in the fact that this human is destined to be a very different creature from all other creatures on Earth. He enjoys free will, the ability to choose, and abilities to perform wilful acts. All this is to test this creature and see how he will use his resources and how he will behave. In Surah Al-Mulk:

> *Who created death and life that He might try you—which of you is best in deeds? And He is the Mighty, the Forgiving.(67:02)*

Such a test requires that there should be an extension of life through an afterlife with either reward or punishment, and that is what the Qur'an exactly explains. However, these explanations may be metaphors for reward and punishment, since the descriptions given in the Qur'an cannot be comprehended within our physical and mental scope. Therefore, we cannot take the descriptions given literally. Furthermore, the terms reward and punishment, are far from being comprehensible, since they are part of a metaphysical state that humans are promised in the afterlife. Consequently, we say in the new *kalām* that we should accept these theological narratives as they are, without too much inconclusive speculation. In short, the duty of man is well defined in Islam: to do good and not be an infidel by doing mischief or being a polytheist. In Surah Al-Kahf:

Say, I am but a man like yourselves; but it is revealed to me that your God is One God. So let him who hopes to meet his Lord, do good deeds, and let him join no one in the worship of his Lord. (18:110)

The scientific explanation for the rare chances that allow for our existence is fundamentally different from the theological explanation given by the Qur'an. Although the Qur'an calls upon us to contemplate creation and to search for its origins, we find that it is primarily concerned with pronouncing the greatness of the Creator, His uniqueness, His oneness, and His dominion over all of existence. The presentation of the view given in the Qur'an is a very good approach in exposing how the methodology of natural sciences differs from the methodology of religions. However, we should admit that the methods and goals of science ignore considering questions which are no less important than questions about our existence; questions which are related to our destiny. But we should also admit that a purposeless universe, acting in accordance with certain rules or laws of unknown origin and drive to produce all this diversity and complex organisation, is something that raises more questions than it resolves. Perhaps, it is because our mental capabilities are not yet equipped to grasp such paradoxes, or that the messages from the Divine were sent to us in order to provide a guide to help us seek our quest in a correct and more efficient way.

IS THERE A PURPOSE FOR THE EXISTENCE OF HUMANS?

Believers in the strong anthropic principle are still unsure of the supreme goal and main purpose for the existence of humans, even though they know the truth behind the anthropic principle and *its* purpose. The absence of a great purpose for the existence of humans on Earth would make the anthropic principle no more than a mere coincidence. Humans, from a materialistic point of view, are part of the material universe; no more, no less. So, why should the universe have been created to serve them? On the other hand, the validity of creating this universe for the service of humans would make the approaches of some scientists, such as John Wheeler, acceptable, if somewhat bizarre. Wheeler states that, "if this is just the human being then what would the universe look like?"

This approach leads to one conclusion: that there is an inevitable purpose for the existence of man. Others, such as John Barrow and Frank Tipler, adopted the same approach, maintaining that, "Our presence imposes serious limitations on the proportion of photons to protons in the universe."[408] These statements of Wheeler, Barrow and Tippler are vague and mean one thing

408 Barrow and Tipler, *Anthropic Cosmological Principle.*

only; that there is an inevitable purpose for the existence of human beings. Now, my question comes in this context: if the human being, according to the materialistic perspective, is nothing more than an insignificant part of this enormous universe, and if this mass of meat and bone, which lives on Earth for a few decades, will only end up as a pile of ash, then what is the ultimate goal of man's existence?

The supreme consciousness and the advanced mental capabilities of humans cannot be just wasteful development, be it a result of natural Darwinian evolution or a careful design and creation. Therefore, the purpose of having humans on Earth cannot be comprehended through the materialistic and physical approach only. Like it or not, we need to have some sort of metaphysical assumptions, otherwise our existence cannot be explained. Human consciousness is a highly valued channel that may work as a link between two worlds: the physical world and the metaphysical world that is beyond our direct senses. It is important, thus, to recognise the value of human consciousness and to distinguish it from the inherent consciousness found in other creatures.

The Islamic understanding of the anthropic principle poses an integrated sense, starting with Allah's willingness to create and ending with a world that accommodates a creature who appreciates its metaphoric value and, consequently, would appreciate the value of creation. Although we do not know the motives of the Creator, we may be able to understand them through investigating creation. Only then may we be able to probe the 'mind' of God. The Mutakallimūn tried to address the question of God's motive for creating the world and came to the conclusion that He could not be driven by a motive because if such a need existed, God would be lacking the self-sufficiency which is necessary for His omnipotence.

Humans have free will (at least in the apparent state of affairs) and limited abilities that they use within their physical and moral activities in order to achieve the aims of their existence on Earth. These aims and goals include exploring the world, discovering how it developed, its movements, and how it works. This is something humans have done ever since they became aware of themselves and the world. But what is the point of exploring the world, and what is the value and usefulness of the knowledge that is obtained?

The first goal is clear; humans invest this knowledge into improving their living conditions and cultivating their surroundings in order to overcome the difficulties of life. For example, when humans discovered fire, they used it to cook their food. Yet, what is the purpose of cultivation and welfare? There must be a higher purpose, and for this we need to view the world from a comprehensive and broader perspective. Was the goal of creating this world, with its atoms and galaxies, and humans—with their extraordinary consciousness and marvellous mental ability—only for the cultivation and wellbeing of all lifeforms? What is the point of this goal if the individual lives only for a few decades before dying and its existence is completely obliterated? Certainly, our minds provoke us to think of a higher and much more sophisticated purpose; a purpose with a deeper meaning and a broader impact than what I have already mentioned. To achieve the sublimity of this purpose and to be consistent with the sublimity of the world's creation, this purpose must be connected to the sustainability of existence, which means the sustainability of human consciousness. This is the result of man's existence in this physical world. If human beings were to understand creation and discover its secrets, then they should be worthy enough to comprehend that there is a greater power who created them and created a world which is subject to them. This fact will not be grasped directly by humans through solving differential equations or by the synthesis of DNA in living cells, but it will be indirectly grasped when they raise their consciousness and add another dimension, a metaphysical one, to it.

I believe that this transcending status can be obtained through intuition and not necessarily through rational reasoning, since the unseen is imperceptible when directly applying the mind, which derives extrapolations using its intuitive knowledge. If one does this *and* invests his partial will and abilities well, then he will know Allah, the Creator, the Omnipotent, and realise that his presence on Earth and the creation of the universe is not, and cannot, be in vain. Therefore, when man gets to know his Creator, he may achieve the ultimate goal of existence; that is to achieve the indefinite sustainability of their existence. Thus, we can understand the anthropic principle in Islam in accordance with its integrated context, starting with Divine will, which is fulfilled at all stages of the universe's creation, to provide conditions for man's existence to implement God's will according to His Divine norm, in which creation is shaped as one entity in all its multiple shapes. All this is intended for humans to know their Lord.

Thus, the anthropic principle, as understood in Islam, is part of a comprehensive and integrated system concerned with the issue of the beginning, means and purpose of existence. The basis of this purpose is to know Allah and to reach Him. For that to happen there must be a path for man to follow. In Surah Al-Ma'idah:

> *O you who believe, keep your duty to Allah, and seek means of nearness to Him, and strive hard in His way that you may be successful. (5:35)*

That may explain the reasoning behind man possessing a limited will and ability, represented by his mental and physical capabilities, in order to achieve the purpose of his creation and development. It is to get to know the Creator in a process of cosmic development and selection that we will ever ascend to until we become part of him.

The relationship of the universal constants, both amongst themselves and in terms of their building the cosmic structure and its parts (large and small), according to the rules and principles that tolerate the presence of other alternatives as well, is not an isolated case. For example, carbon formation is essential for the existence of life, but it is not the only requirement. It is possible to have life forms without carbon, as we mentioned earlier, and scientists have developed pathways for life forms to exist, which do not depend on carbon for their emergence and development. These forms of life will, inevitably, be different from ours, although the aim of creation could remain unchanged and the intermediary of creation (the successor) remains as it is, too. In addition, the anthropic principle also remains as an intermediary and a guide towards Allah. Hence, we can understand the meaning of the verse in Surah Fatir:

> *If He wills, He could dispose with you and bring forth a new creation. (35:16)*

Therefore, Muslims are not confused when it comes to the anthropic principle and do not get lost in the dilemma of whether the universe has been created for the sake of humans or whether *they* exist for the sake of the universe and its knowledge. Since Muslims know that humans are rational beings who were created to serve a purpose, it would be insane to think that these rational beings could be servants to an irrational object. The Holy Qur'an has shed light in many places on these cases. We quote from Surah Yunus:

> *They worship beside Allah that which neither hurteth them nor profiteth them, and they say: These are our intercessors with Allah. Say: Would ye inform Allah of (something) that He knoweth not in the heavens or in the earth? Praised be He and High Exalted above all that ye associate (with Him)! (10:18).*

When it comes to parallel worlds, I could not find anything in the Qur'ān that argues against, or completely negates, the idea of the existence of other worlds. There is no theological restriction on the number of worlds which can be created by Allah, as long as this number is finite. The same applies to the finite number of atoms in the world. On defining finiteness, Muslim scholars relied on the following verse from Surah Al-Jin to verify their arguments:

> *He (Allah) keeps count of all things (*i.e. *He knows the exact number of everything). (72:28).*

To conclude, I should emphasise that Allah rules the world through well-defined blind laws, which are formatted to be indeterministic in order to allow for his intervention by choosing which of the possible events might affect us. Blind laws are necessary to establish causal relationships, which are themselves necessary to establish law and order in the world. It is, then, a basic requirement to recognise the Creator and Sustainer of the world. This vision surely needs more involvement with other issues related to Divine action and could be discussed in the new *Jalīl al-kalām*.

Chapter Thirteen

Evolution

The evolution of living organisms is one of the most important of contemporary issues. Its significance is manifest in some major respects, which are primarily related to the understanding of the constitutive realities of living beings, their interactions, modalities, and the inter-relationships of these vital formations. This is necessary if we are to comprehend the mechanics of their development and the reason for so much of the diversity we see. This comprehension enables us to learn how to deal with any emergent changes that may occur in their morphology or physiology. In addition, it would facilitate an understanding of any such deficiencies present which, in time, might manifest themselves as diseases and deformities.

THE MYTH AND BELIEFS

Texts from the monotheistic scriptures provide us with a particular vision for the existence of mankind on Earth. They tell us that the first man was created by God out of mud moulded into the shape that we see and blessed with the Spirit of God, which transformed him into a human.

There have been a wide range of explanations for these religious texts. One of them was the literal embodied perception which the Jews, in particular, implemented in their interpretation of the sacred texts. They believed that God made the first man out of wet mud, fashioned him, and then breathed life into his body. However, some misunderstanding emerged when they interpreted the words, 'God created man in his own image', from which they understood that man was created in God's image. Islamic exegeses is replete with myths about the emergence of the universe and its life source and are mainly inspired by the ancients and their tales of the creation. In this respect, perhaps most of Islamic written heritage bears a similarity to the creation myths of the Old Testament and legends regarding the separation of sky from water and the emergence of the Earth, as described in Genesis.

In the legacy of Islam, we can spot some of these tales (some of which were attributed to the Prophet Muhammad (peace be upon him), while others are narrations of his companions, such as Abū Hurayrah, Ibn ʿAbbās, and Kaʿb Al-Aḥbār). In addition there are the esoteric explanations that have burnished the story of creation as symbolic parables rather than being subject to changeable interpretations depending on variant signs and other symbols. The quest for the origins of life and its manifestation on Earth, is not only a challenge for the human mind but also an adventure that might lead to a confrontation with our traditional understanding of some of the principles of faith that we have inherited from the culture of our ancestors. This question has become one of the major contemporary philosophical issues, and in this chapter, I will present some of the challenges and intellectual problems that are related to the issue of creation of man and his subsequent evolution over the ages.

SCIENTIFIC CONSIDERATION

Man has been unable to explore the depths of this issue in a serious manner until recently, when he developed the theoretical, functional, and exploratory tools which enabled him to shed light on the virtual matter relating to the beginning of creation. This was investigated by the English zoologist, Charles Darwin, and many of his predecessors and contemporaries. In the first half of the 19th century, Darwin recorded a large number of observations and documented the living organisms which he collected during a voyage which covered much of the southern hemisphere.

After studying these observations and documents, Darwin came to the conclusion that organisms have evolved biologically through the ages. The shapes and function of their organs, in addition to their capabilities, change according to the requirements of their surrounding environmental conditions. This is known as adaptation. If they cannot adapt, creatures would not be able to live and would become extinct. An example is two types of butterflies; one of which can change its colours to be compatible with the environment in which it lives, while the other cannot. Adapting provides the first butterfly with the ability to hide and camouflage itself from predators that might otherwise pounce upon it and eat it. The other type of butterfly, meanwhile, does not have the ability to change its colour and, being clearly visible, will be easily attacked by predators, generation after generation. Using this scenario, obviously, the second type will become extinct, whereas the first will prevail due to its camouflage and hiding ability.

According to Darwin's original theory, the evolution of living beings is a sequence of random events that happens to show new traits in subsequent generations. These traits, when challenged by the environment, will be tested and result in either success or failure. If it fails to withstand the impact of the environment, the creature will perish. If it withstands the requirements of the environment, the evolving creature will survive. This interaction with the environment is called Natural Selection.

Whenever an organism encounters adverse conditions, the evolution of its physical body and physiological functions through subsequent generations will ensure its survival. If there is a positive development which eventually helps the organism to counter what was once invincible, then this kind will survive but if the new development lacks the required characteristics, then this species will vanish. Hence, the term "survival of the fittest," which arose from this principle.

Following this, Darwin decided to include all the patterns of life in a full scope of evolutionary biology. He claimed that all organisms have one origin, from which other living species have been evolving as a series of ramified patterns; arising from one another as determined by genetic mutations. The structure of genetic heredity, discovered in the 1950s by Watson and Crick, allowed for a leap in understanding the dynamical structure of evolving, living organisms and formed a basis for the accurate identification of the link between organisms. This discovery, in particular, supported Darwin's theory about the origin of species and paved the way for Neo-Darwinism.

Darwin's theory of biological evolution, cast into a modern understanding, is based mainly on the assumption that there are two mechanisms at work. The first mechanism is that of *random genetic mutations* taking place in chromosomes. The second is *natural selection,* by which nature is thought to choose the 'fittest' of the offspring for survival, while the weak perish. In other words, if the new qualities are unable to cope with the surrounding environmental realities, this will highlight the weakness of the living organism and, thus, result in extinction. Hence, the fittest organisms are adapted solely to survive on Earth, which highlights the notion of natural selection. As long as the struggle for survival is an ongoing process, the organic evolution of living organisms will persist.

THE VIEW IN THE NEW KALĀM

The question of organic evolution is one of the emerging issues that should be addressed within the framework of the new *kalām*. Therefore, I will present, in this chapter, an analysis of the question of biological evolution from both the theological point of view and as a scientific analytical argument.

Qur'an and the Evolution of Living Organisms

The first question that arises is whether or not biological evolution is in agreement with the Qur'an. It is commonly believed that the idea of organic evolution of man and living creatures is incompatible with religion. Relying on a literal understanding of religious scriptures, people have so long supposed that living organisms were originally and directly created by God out of soil, and that mankind was created from clay, which was then moulded into the convenient form of the human being. Until recently, lay people believed that insects, worms, and ants self-arose from non-living materials; a point of view formerly adopted by Aristotle and many of the Elders of Greece.

Several verses of the Qur'an imply that God created man from clay (or dust). In Surah Sa'ad:

> *(Remember) when your Lord said to the angels, "Truly, I am going to create man from clay." (38:71)*

And Surah Al-Imran:

> *Verily, the likeness of 'Iesa (Jesus) before Allah is the likeness of Adam. He created him from dust, then (He) said to him, "Be!"—and he was (03:59)*

> *O mankind! if ye are in doubt concerning the Resurrection, then lo! We have created you from soil, then from a drop of seed, then from a clot, then from a little lump of flesh shapely and shapeless, that We may make (it) clear for you. (22:05)*

> *And of His signs is this: He created you of soil, and behold you human beings, ranging widely! (30:20)*

> *Allah created you from dust, then from a little sperm, then He made you pairs (the male and female). No female beareth or bringeth forth save with His knowledge. (35:11)*

The seventh verse of Surah Al-ʿImran explicitly states that the Qur'an contains verses which are firm and clear, whilst others are allegorical or difficult to interpret (*mutashābihāt*). Most of the verses of the Qur'an have a straightforward meaning when they handle issues related to sacred orders, religious orders and Sharīʿa laws. On the other hand, the verses which are allegorical are those covering topics concerning cosmological arguments, such as the creation of man and the universe, or describing metaphysical matters of creed. As a consequence, this paved the way for exegetes to present their ideas and interpretations. However, some people extracted meanings which, perhaps, deviated from the intended meaning by polishing them with contrary indications. Yet, the correct interpretation of Qur'anic texts abide with the context and grammatical rules of the Arabic in the first place. The Qur'ān asserts this by stating:

We have revealed it—an Arabic Qur'an—that you may understand. (12:02)

The reason why there are problematic verses in the Qur'an is explained by the fact that the Qur'an expresses the word of God in classical Arabic. Many of the sentences are structured in such a way as to hold a variety of clues, many of which are not interpretable by man, who has a limited cognition compared to that which the Almighty propagated in the Qur'an. The more man's acquaintance with the Qur'an blossoms with time, the more meanings will be exposed. As a result, the Holy Qur'an has been, and will always be, a perpetual truth.

It makes sense to say that it was not possible to disclose the contents concerning creation and other sophisticated issues all in one go to a people who would have had great difficulty in comprehending them and, hence, accepting them at certain periods of time. Had the Qur'an revealed these connotations, it would have required a lot of explanation and induction to cover many other fields of science, which might not necessarily have ensured a proper comprehension for the matter involved. Therefore, the verses concerned with these subjects have maintained an ambiguity in which the general meaning dominates, and which is bare of further details. The creation of the seven heavens, for example, and their destiny is open to much potential interpretation and the reader has the choice of figuring out the intended implications as long as the general meaning is clear.

Another motive that impedes many people from accepting the idea of the organic evolution of living organisms and humans is their fear of diminishing the solemn role of the Creator if they recognise the mechanism of natural evolution and, consequently, being accused of atheism. On the contrary, the unequivocal truth prescribes the following purview: that contemplation, not only of the correlation that knits the universe with the Creator but also of the natural mechanisms running the world (which seem to be axiomatic and spontaneous), confirms the unquestionable truth that Allah is the Living, the Self-Subsisting, and the Eternal, who has power over everything.

As I have shown in previous chapters, the laws of nature, which guide the natural mechanisms of the universe, are manifested with probabilistic characters rather than deterministic ones. This was uncovered by quantum theory, which is one of the pillars of modern physics. This is also exemplified in other fields of science, such as the laws of chemistry and the life sciences which employ atomic and molecular reactions to explain biological phenomena. Therefore, the laws of biochemistry and ecological transformations do not hold deterministic results. On the contrary, these processes are based on probabilistic outcomes, a fact that has been proven by modern science and the discoveries of the 20th century.

Once we recognise that biological operations have a large number of possibilities, it is natural for us to wonder what controls those possibilities. It is necessary to realise that no single molecule of the reactants can be thought to assume either control of the driven operation or the choice of the probabilistic results. Such a controller, who dictates a specific choice of occurrence of an event with a low probability, must be knowledgeable of all the parts and their properties, and be familiar with their conditions and requirements. And he has to be supreme with respect to all choices and purposes so that a positive evolutionary result is obtained. This reminds me of what the British physicist, Paul Davies, quoting the famous astrophysicist Fred Hoyle, said, "A common sense interpretation of the facts suggest that super intellect has monkeyed with physics, as well as chemistry and biology and that there are no blind forces worth speaking about in nature."[409]

Does the Qur'an Oppose Biological Evolution?

The answer to this question has two levels: the evolution of man and the evolution of the rest of creation. With regards to the second level, the Qur'an has mentioned nothing about the rest of the creation, except telling us that:

> *And there is no animal in the earth, nor a bird that flies with its two wings, but (they are) communities like yourselves. (06:38)*

Below, I will present my understanding of the Qur'anic texts with reference to creation, with a purpose to investigating the truth as presented in the Qur'an. The methodology of this research is based on reviewing those Qur'anic texts concerning creation, and the emergence and evolution of man, and taking into consideration the Arabic language, in addition to the many interpretations accredited to Muslims, one of which is that of Ibn Kathīr, so as to be familiar with what previous scholars believed.

The main concern of the Qur'an in narrating the story of the creation and evolution of man was to specify that this event was ordered by Allah according to His will. We understand that Allah informed the angels of His will to do this but that the angels were confused in their understanding. We read:

> *And when thy Lord said to the angels, I am going to place a ruler (a placement) in the earth, they said: "Wilt Thou place in it such as makes mischief in it and*

409 Davies, *The Accidental Universe*, 118.

sheds blood? And we celebrate Thy praises and extol Thy holiness." He said: "Surely I know what you know not." (02:30)

It is not clear why the angels questioned the will of God, but it is clear that they suspected this 'being' as someone who might do mischief and cause bloodshed. This indicates that they had a previous experience with such a creature. This can only mean the available creatures at the time, i.e. the animals and such like.

In another verse the Qur'an tells us:

(Remember) when your Lord said to the angels, "Truly, I am going to create man from clay; So when I have made him complete and breathed into him of My Spirit, fall down submitting to him." And the angels submitted, all of them." (38:71-73)

This verse, taken in isolation, may give us the impression that God brought some mud and moulded it into the form of man, then breathed into him of His Spirit, so that he became a live being, and that the angels then submitted to this new creature. However, if we investigate the other verses of the Qur'an concerned with this topic, we see that there are staggering details that should not be overlooked. These details explain the stages by which man evolved to be a creature deserving the submittal of the angels. The Qur'an tells us that God *started* creating man from clay.

...and He began the creation of man from clay. (32:07)

But what sort of clay was that? The Qur'an tells us that it was wet, rotten clay. For this, there are three verses in the same Surah, al-Ḥijr, where we read

And surely, We created man of sounding clay, of black mud fashioned into shape. (15:26)

And when thy Lord said to the angels: I am going to create a mortal of sounding clay, of black mud fashioned into shape. (15:28)

But the Shaytan (the devil) objected saying:

I am not going to make obeisance to a mortal, whom Thou has created of sounding clay, of black mud fashioned into shape. (15:33)

In the above three verses we spot the words 'sounding clay' (in some translations it is sounding pottery), then we spot the words 'black mud', which indicates a mineral-rich, old and decomposing mud. This understanding is well supported by texts from the original exegesis of the Qur'an. Therefore, we can consider it reliable enough to express the meaning of the verse. Accordingly, we conclude that God began to create man from decomposing mud. But where can such mud be found, except in ponds? Incidentally, the exact words of the verse say that man was created from *ḥama' masnoon*, the word *ḥama'* meaning old mud, while *masnoon* could imply several meanings. One is that it is old and has been there for many years. The other, which I find more meaningful, is to say that it means old mud which has been assimilated for the intended purpose, which is the creation of man.

The question then arises as to whether the creation of the other creatures was from the same mud or not. There is no indication in the Qur'an for this, but experience tells us that the style of the Qur'an, when presenting a topic, focuses on the main issue rather than going into details of related topics. The topic here is the creation of man and there is nothing about other creatures being created in the same way. However, we should remember that we are dealing here only with the beginning, i.e. the first stage.

In the second stage of creation, the Qur'an tells us about the preparation and development of man. This, we can understand from several verses which mention the care taken in making man. We read:

> *So when I have made him complete and breathed into him of My Spirit, fall down making obeisance to him. (15:29)*

> *So when I have made him complete and breathed into him of My Spirit, fall down submitting to him. (38:72)*

> *Who created you, fashioned you perfectly, and enabled you to be upright. (82:07)*

Many of the available translations of the Qur'an do not give the correct translation of this verse, (82:07), maybe because they do not like to imply any indication of evolution in the ascent of man, or they are just unaware of this implication.

A narration of the Prophet was documented by Ibn Kathīr, indicating that the fashioning stage correlates with the ability of man to walk upright on his feet, which is considered to be one of the stages of human evolution. The Prophet said, "God says: "O, the son of man, how would you challenge me while I have created you out of this [spit], and once I fashioned you upright you walked proudly collecting [wealth] meanly and once you feel dying, you say now I will be charitable and what a time for charity." This important Ḥadīth Qudsī is widely held as being authentic.[410]

The third stage of man's development was breathing into him with the Spirit of God. At this moment, man became human through acquiring partial will and intellect. This enabled him to choose, invent, create, and develop things for his benefit or, unknowingly, his misery. At this stage, God ordered the angels to prostrate themselves before the human being, indicating that

410 Ahmad Ibn Ḥanbal, *The Musnad*, Risala Edition, vol. 29, 385.

they, along with the rest of the world, would be subservient to this creature. At this third stage the man evolved to become human by obtaining cognitive abilities. At this stage, the Qur'an tells us that Allah taught the new creature, i.e. Adam, all the names of things as an indication of having acquired cognitive capabilities:

> *And He taught Adam all the names, then presented them to the angels; He said: Tell Me the names of those if you are right. (02:31)*

Man's subsequent story, as told in the Qur'an, informs us that the propagation of the species was accomplished through mating between the sperm of the males and the eggs of the females, referred to as 'fluids despised.' Here, we see in the Qur'an two verses which appear to have two different meanings, the first saying:

> *Then He made his offspring from semen of worthless water (male and female sexual discharge). (32:08)*

While the other verse says:

> *Did We not create you from a worthless water? (77:20)*

In my opinion, the first verse (32:08) is addressing the creature that was developed prior to the formation of the first human being (Adam, so to speak). The evidence is the context itself, which intimates that the creature has not yet received the Holy breath. Verse 32:08 continues:

> *Then He made his offspring from semen of worthless water (male and female sexual discharge). Then He fashioned him in due proportion and breathed into him the soul (created by Allah for that person), and He gave you hearing (ears), sight (eyes) and hearts. Little is the thanks you give. (32:08–09)*

Again, in order to avoid indicating that there are two (or more) different creatures among the origins of human descent, most exegeses of the Qur'an choose to interpret this verse such that they view the Holy breath to be an event which occurs at a certain stage of the foetus' development in the womb. However, this interpretation is in conflict with the phraseology of the verse which mentions the 'offspring' (*naslahu*), which is not a foetus but a newborn child.

Furthermore, breathing the Holy Spirit into the body, as presented in the third stage of creating a human being, was a turning point in the evolution process. The creature was transformed from the uncivilised stage, where he had no developed mind, into a stage where he could think and deduce using an advanced intelligence. This, I believe, was imparted by Allah through His Holy Spirit. That breath altered man from being a higher animal into a human being. Following this stage, the angels (which is a metaphor representing the laws of nature here) were ordered to be at the disposal of humankind, to enable him to explore the world, discover it and exploit it positively for his own welfare. Through this, man was supposed to acknowledge the greatness of his Creator, glorify Him, and become part of His high-ranking kingdom.

From the interpretation given by Ibn Faris in his lexicon[411] of Arabic, where he provides the origin of words out of their literal construct, I understand that the bowing of the angels before man was meant to put them at the disposal of humankind. Of the word bow (*sajada*), he says that its origins point to lowness and submissiveness. It is commonly known that a low-ranked subject prostrates before the higher-ranking person. The reason for the angels bowing revolves around the fact that they are Allah's agents, His messengers to the world, and those by whom Allah orders the world. In other words, the angels prostrated before Adam because he was given priority and privilege above them, and he is also superior to them due to his strength, firstly, and his deeds, secondly.

Based on this presentation and analysis of Qur'anic verses, it may be concluded that the creation and development of humankind does not necessarily contradict the possibility of having some sort of biological evolution through which humankind has developed to its present stage. In terms of creation and evolution, it is emphasised in the Holy Qur'an that all that happens is predestined by Allah, and takes place according to His will and the laws of logic. Additionally, there is a possibility for having an evolutionary setback or a developmental change, resulting in a malformed creature; a metamorphosis, in other words.

Here, I am not claiming that our understanding of the Qur'an image is a comprehensive one. On the contrary, there must be an unknown, metaphysical aspect to the image which involves the creation of Adam in the Garden. It cannot be a clear-cut conclusion that the Garden is on Earth, although some scholars consider that the issue of being removed from the Garden and sent down to Earth was only meant to serve a spiritual purpose. Ultimately, the unintelligible passages in the Holy Qur'an are in need of further analysis in order to come up with solutions and decipher their mysteries.

Another significant point, which might be thought to conflict with religious belief, is the claim that man originated as an animal: a monkey, deer, fish, or some other creature. Some religious doctrines (not necessarily in the Qur'ān) confirm the idea that these creatures were the result of metamorphoses which do not conform to the concept of honouring man and making him a vicegerent on Earth by the Creator. But this problem does not arise in the Qur'an.

---

411 Ibn Faris, *Mujam Maqayīs al-Lugha*, edited by Abdul Salam Harūn, Dar al-Fikr, Beruit, 1979, entry: sajada.

The honouring of mankind as the sons of Adam came after the advanced creature became Adam-like. The primitive creature became a human being when the Creator fashioned him, straightened him and blew the Holy Spirit into him. The morphological, anatomical, physiological, and even behavioural facts indicate a similarity between the species, known as *Homo sapiens*, and the other advanced creatures. We, as humans, eat, drink, breed, and practise many of our daily life activities in a way which resembles those of other animals. The difference between us and other high-ranking animals is simply that we have the skill and constructive logic by which we are able to comprehend the world. Without this talent, we might descend into the lower animal class. Should we choose to ignore that not only is there a purpose behind our existence, but there is also a power, a will and an objective for the world in which we live, then we will resemble other members of the animal kingdom to which we originally belonged. This means that our class is spiritually advanced compared to the animal kingdom inasmuch as we have the ability to distinguish, innovate and construct. Yet, if we underestimate this gift, along with the facts to which it guides us, we will become like cattle or even worse. Unlike humans, animals do not have rational faith as they are not qualified for this. The Creator raised man over other creatures to enable him to meditate and think, so as to be the vicegerent who innovates and perceives the universe using his senses (although he has grasped only a small portion of the physical part of the universe). Accordingly, a human being has to appreciate these values and respect them at all levels and in all activities.

Scientific Analysis

One can confirm that the notion of biological evolution of living creatures is a good explanation for numerous observational facts. This makes it a basic pillar in modern life sciences. It would be difficult to recognise modern biology as a major field of science if our life were devoid of this principle. Those who deny biological evolution should submit a legitimate rationalisation for the emergence of species on Earth. They should also be capable of justifying the morphological, physiological, and structural similarities, in addition to the innate sociological interaction between themselves and other creatures of the animal kingdom. Moreover, they would have to explain the success of those scientists who adopted biological evolution in their research, as this success indicates a correct approach that has been followed for many decades now.

However, we should be aware of the claim that biological evolution does not necessarily imply the endorsement of the neo-Darwinian approach for evolution. It does not mean that all theoretical explanations proposed by the neo-Darwinians are correct, as scientific evidence suggests that some theoretical details are still unknown. Consequently, a distinction should be drawn between acknowledging biological evolution, on the one hand, and believing in all the theories of evolution, on the other. The former demonstrates an established reality which is backed up by a great deal of evidence, whereas the latter is controversial and should not be taken for granted.

Some authors, such as the Frenchman, Jean Staune,[412] consider the theory of evolution as being incomplete and observe that it is somewhat similar to the theories of motion set during the times of Galileo and before the discovery of Newton's laws. This means that the current theory of biological evolution needs to undergo many transformations in order to proffer the reality of evolution as a complete sequence. For example, claiming that evolution rests on two pillars (a claim adopted by the neo-Darwinists): random mutations and natural selection are incompatible with the fruitful results of evolution that we see occurring in nature.

Darwinism believes that the division of sexual cells can result in random mutations within the genetic material. Consequently, these mutations, which emerge during the copying process, cause changes in the hereditary code, and their effects may become manifest during the development of the organism. During its lifetime, a conflict for survival is perpetually taking place between the creature and its natural habitat. Only when the result of this conflict is known will it be shown whether or not this biological mutation is of benefit. If the mutation is a beneficial one, this will lead to the survival of the creature and give it a better opportunity of breeding than others whose DNA did not incorporate the mutation. On the other hand, if the mutation results in a severe impairment, then the creature may die, as might other organisms with equally non-favourable mutations. Since nature dictates whether or not creatures with developmental mutations are to survive and reproduce, the selection process has been bestowed the term of Natural Selection. Logically, the creatures carrying the preferred mutations will breed and be prosperous, while those carrying detrimental mutations will decrease and become extinct. This is an illustration of the mechanism suggested by Darwin's Theory of Evolution, according to modern understanding.

From the Islamic point of view, biological evolution, as I have shown, is acceptable and can easily be accommodated within the framework of the Islamic worldview, provided an accurate interpretation of the Qur'an is made available. In fact, I don't see a major inconsistency between Darwinism and Islamic belief, except when assuming genetic mutations to be totally random. If one believes that mutations are entirely random, then it implies that they are independent of the Creator's will. Obviously, what seems random is not necessarily random, if we consider the previous chapters and the actions of

412 Staune, J. "Does our existence have a meaning?" (Paris: Presses de la Renaissance, 2007).

the innate laws of nature. The fact that the results of these laws are probability based rather than deterministic negates the notion of randomness and asserts the principle of indeterminism.

Necessity and Contingency in Evolution

There are some serious loopholes in the neo-Darwinian explanation of evolution. Alternative non-Darwinian approaches have been suggested.[413] I do not want to get involved, here, in the details of such criticism but will stress again that the random mutation-natural selection process cannot be accepted as a mechanism for generating new, fruitful and ascending evolution as we observe it in nature. It cannot guarantee the efficient diversity of creatures on Earth. Random mutations mean that equally probable events may, over a long time, cancel out and produce nothing, despite the alleged natural selection factor.

The mechanism of evolution might be understood through the existence of two necessities and many contingencies. The first necessity is the need to adapt to environmental conditions. For this purpose, new features in the morphology or the physiology of the creature are essential. On the other hand, the blind law of nature necessitates both a driver and a coordinator in order to guarantee a harmonious and fruitful output. For each of these necessities there are many contingencies, perhaps endless possibilities, but there will always be only one optimised solution. Such a solution picks those contingencies which will, ultimately, achieve maximum efficiency and harmony. However, since the laws of nature are probabilistic, there will always be a few choices of unoptimised contingencies. Potentially, this would cause a failure, with evolution producing undesired traits. This vision may provide a basis for a mechanism of evolution that satisfies better observational facts.

Microscopic Evolution on a Quantum Base

The behaviour of atoms and molecules in chromosomes modulate biochemical reactions, which may result in genetic mutations. The laws of quantum mechanics control this behaviour microscopically. These laws were formulated in the first half of the 20th century as a result of the research of Max Planck, Niels Bohr, Werner Heisenberg, Louis de Broglie, Erwin Schrödinger and many others. The laws had a logic regarding the microscopic behaviour of particles. This logic is somewhat different from the traditional logic of classical physics. The most important features of this logic are those related with probability and determinism: events that happen are probabilistic, i.e. they are singled out of many contingent events that depend on the state of the system in conjunction with the process of measurement. The basic aspects of this theory have

413 King, Jack Lester and Jukes, Thomas H. (1969) "Non-Darwinian Evolution." *Science* 164 (3881): 788–798.

been clarified in Chapter Five of this book, especially the issue of measurement of quantum mechanics, along with different explanations for this theory. I have contributed my own explanation for the problem of quantum measurement, which is based on the notion of re-creation, as borrowed from *Daqīq al-kalām*. Aside from the chronic problem of measurement, quantum theory is the best available framework for understanding the interaction of matter and energy, and the behaviour of elementary particles, atoms and molecules. Since biochemistry is concerned with reactions between these atoms, molecules, and ion transport to form different compounds (the components of biological materials of the living substances) quantum mechanics and logic are undoubtedly involved in figuring out these compounds through the topic of quantum biochemistry.

Johnjoe McFadden, a molecular genetics professor at the University of Surrey, attempted to make use of quantum mechanics and biochemistry in order to work out chemical evolution at the molecular level of basic biological compounds in a protocell. He presented the general framework of his theory in a book entitled, *Quantum Evolution: Life in the Multiverse*,[414] where he said that these quantum activities have an effect on the whole structure of a living creature because a single DNA molecule may affect all the other cells. Since microscopic composition of the molecules are ruled by quantum indeterminism, quantum mechanics undoubtedly has an essential role in composing the amino acids and the bases that constitute genes—the hereditary elements of living creatures.

It is also known that the possible number of protein compounds which can be made up from these amino acids and alkaline bases can run to billions, whereas, in reality, we see only a small number of these compounds formed. This is a real problem which biochemistry faces at the molecular level. In order to clarify the size and importance of this problem, McFadden undertook the construction of peptides as a result of the interaction between twenty alkaline bases and thirty-two amino acids. The number of probable compounds, based on random distribution, was estimated to be as many as between 20^{32} and 10^{41} compounds, which is a very large number indeed. If we hypothesise that the random constructions are going to make only one molecule of each of the peptides, then this will require a lot of carbon, much more than is available in all the trees on Earth. Ideally, there should be a rule which governs these probable selections and greatly decreases the available number of peptide chains. For this purpose, McFadden utilised the superposition of states principle, a fundamental pillar of quantum mechanism, as postulated by the Copenhagen School. This explanation says that the quantum physical system, in the absence

414 Johnjoe McFadden, *Quantum Evolution*, Harper Collins, 2000.

of measurement, is a superposed status of all possible states. At the moment of measurement, the physical state of a system collapses on one of the possible states in what is called wave-function collapse, or the symmetry between the observer and what was assigned on the scene, as previously mentioned. Briefly, McFadden devoted a great deal of effort to justifying the availability of the quantum states in limited peptide chains, which can breed automatically. He explained this descriptively without providing any detailed calculations.

However, McFadden encounters a major problem in terms of quantum measurement when he asks, "Where have other peptide chains disappeared?."[415] Here, McFadden employs the idea of multiverse, hypothesising the existence of 20^{32} worlds with our universe being the only one to have acquired the peptide that is capable of breeding automatically. But McFadden immediately understands that the multiverse idea necessitates faith that what has happened in one universe will not happen in other multi universes. Were it applicable to the universe we live in, it would be a unique event which cannot be duplicated. In other words, McFadden suggests the scenario, which he formed to develop the peptide which can breed automatically on Earth, as one which cannot recur in any other place in our universe. This means that life in our universe can only be expected to be on Earth. Based on the same argument from the multiverse interplay, McFadden concludes that it is impossible to repeat the development of life in a laboratory experiment. He believes that, even if scholars gather all required conditions to establish life in a laboratory, they will not be able to regenerate life. This belief reminds me of what the physicist, Lee Smolin, said that the multiverse hypothesis can neither be proved, nor can it be disproved.

In summary, it is clear that McFadden's reliance on the Copenhagen interpretation of quantum measurements, in addition to exploiting the notion of the multiverse in his attempt to present a scenario of producing self-replicating peptides, was not adequate enough to prove the validity of the form; there were no detailed calculations in this regard either. Matthew Donald, of the Cavendish Laboratory in Britain, wrote an essay criticising what McFadden presented and describing his method as completely wrong.[416] McFadden and Jim al-Khalili responded to Donald's essay, providing arguments in favour of their scenario. In essence, what McFadden suggested on the process of producing self-replicating peptides is still incomplete, although it contains some interesting points in proposing a major role for quantum theory on the development of the creature at its early stages.

415 McFadden, *Quantum Evolution*, 227.

416 Donald, M. J. Book Review of Quantum Evolution by Mathew J. Donald, Jan 2001, arXiv. quant-ph/0101019.

Difficulties with Neo-Darwinism

As I mentioned before, Neo-Darwinism is based on two principles: random mutations and natural selection. The aim, at this point, is not to discuss the theory of evolution in detail but rather to focus on the general, as well as the fundamental, aspects of this theory. The discussion is supported by a number of general proofs which are characterised by negating false information through verifying the inconsistency in results aside from the procedural details. We may consider that natural selection is actually possible, whereas hypothesising about random mutations is more problematic, due to the fact that randomness cannot escalate fruitfully, as is the case with evolution of living creatures.

Advocates of Darwinian evolution believe that the advancement of living creatures is a result of cooperation between natural selection and useful mutations. However, a simple calculation involving the random options required for the development of a complex organ would reveal the need for a great number of eras (perhaps longer than the age of whole the universe) in order to create a sizeable qualitative development in those organs. This problem may be overcome if there is a steered development. That is to say mutations which are not random but somehow guided by factors from within or without the creature through an, as yet, unknown power.

To say that mutations occur at the behest of some unknown actor might imply that the universe is run by miracles. This, in turn, would invalidate causality and deprive science of the most efficient element for its credibility. Such an invalidation of causal relationships is not acceptable in the new *kalām*. Causal relationships are acknowledged, secondary causes exist but causal determinism is refuted. Undoubtedly, mutations are a fact of life that can be verified in a laboratory but to explain them as occurring randomly involves metaphysical assumptions. It is saying, implicitly, that for some, unknown, reason, mutations are occurring. Describing mutations as 'a mistake' in copying the DNA is again not an adequate portrayal of what could be actually happening. For example, psychological issues, such as stress, may sometimes play a role in producing genetic mutations, while variations in the Earth's magnetic field may have substantial effects in directing the tiny ions involved in the genetic reproduction process. There are many pieces of evidence on physical-psychological effects from various perspectives, such as sudden fright and how it can cause physiological syndromes due to the quick and excessive production of some hormones.

An important piece of experimental work was conducted by the American biologist John Cairns, the findings of which were published in Nature.[417] John

417 Cairns, J., Overbaugh, J., & Miller, S. (1988). "The origin of mutants." *Nature*, *335*(6186),

Cairns isolated a group of bacteria, unable to utilise lactose and grew them on nutrient poor medium such that their numbers decreased dramatically. When he transferred several of these clones onto a lactose rich medium, he observed that some of the bacterial clones were able to grow and produce colonies. These, he subsequently discovered, had acquired genes that enabled them to metabolise lactose and, consequently, survive. The findings of this experiment were considered controversial because there were no known mechanisms explaining how genetic development could be directed such that it would produce the genes necessary for the break down of sugars. The known evolutionary mechanism of genetic mutations stated that the route of information was from DNA through RNA to the protein and not vice versa. Until recently, there has been no known biological mechanism which enables the movements of information from the environment to the genes to instigate their division. This is the problem.

Johnjoe McFadden published a paper in collaboration with Jim al-Khalīlī, which considered adaptive genetic mutations in terms of quantum mechanics.[418] Although, after 1988, many research papers were published which supported the idea of the occurrence of adaptive mutations, they did not provide a viable explanation. However, does this lack of an explanation for certain phenomena allow us an excuse to reject it?

The advocates of Darwin's evolutionary theory emphasise that all life originated from one cell and, therefore, that all the species of all living creatures have one single point of origin. It is the reason why the length of time required to bring about these variations in species is under scrutiny, and not decisively definite. It may be speculated, however, that it might be a very long time, extending perhaps beyond the age of the Earth. This is another argument against the claim for random mutations.

Let us have another example; suppose we have a large number of similar tools made of the same material but in different shapes (household cutlery, for example) and we ask someone to work out the relationship between these tools. He might say they are made by one person, and if not by one person, then at least the source of knowledge is common for all manufacturers. If we tell him that we do not have convincing evidence for the existence of this one manufacturer or those common knowledge manufacturers, he might conclude that then these tools may well have made themselves. Then, if we might ask him whether each tool makes another one or are the tools generated from each other? Of course, the simplest reply will be that they generate spontaneously,

---

142-145.

418 McFadden, J. and Al-Khalili, J. A quantum Mechanical Model of Adaptive Mutation, BioSystems 50 (1999) 203–211.

that they must have developed from each other according to the needs they encountered. If we tell him that these tools do not have intelligence, no hindsight to distinguish or a will, and that they are blindly copying each other, he might say: *aha, they must have developed randomly; a knife changed into a spoon and the spoon into a fork, for instance. If the knife randomly changes into an impractical object, then it will be discarded since it will have no use.* Based on this argument, the same person might say that these devices must have been used for different purposes and that the purpose of the device determines its material and shape. Accordingly, external factors decide the shape of the devices. If the same person finds one of the metal tools containing a plastic part, he will be surprised and ask about the source of plastic. He will question the feasibility of changing metal into plastic. He will get more and more perplexed but remain saying that these tools result from gradual and slow changes from one and the same origin.

This analogy is similar to the arguments put forward by Darwinists from material propagated by Richard Dawkins who denies the existence of the Creator and does not allow himself to think for a moment about the possibility of having a Creator who can direct evolution through certain laws and choose among the probabilities of the possible mutations. Who can falsify that theory of evolving cutlery?

A SUMMARY ON BIOLOGICAL EVOLUTION

The common belief that the evolution of living creatures contradicts religious convictions originates from the concerns of some people that, in accepting evolution, they would have to deny the role of the Creator. They believe that the laws of nature would then, automatically, replace the will and act of God. However, people should now realise that this is not true since the effectiveness of the laws of nature is limited by the probabilities of the results and that these probabilities are not chosen by the laws themselves. Moreover, such a fact uncovers how the laws of nature themselves, including the laws of biological evolution, would be in need of a driver that should not be part of the constituents of the universe.

This does not necessarily mean that the Holy Qur'an supports the evolutionary theory of Darwin or any of the others. The principle of evolution is something different from the theory of evolution. This is because the theory, at hand, has many deficiencies, loopholes, and drawbacks. The idea of the emerging species based on the assumption of completely random mutation cannot be accepted. The evolution is ascending, and it would be hard to know how such a well-organised hierarchy can emerge out of random chances.

As I have said, the theory of evolution is something different from the principle of evolution. As far as I can see, the Holy Qur'an does not object to the notion of the biological evolution of human beings, nor does it set out any specific details for the mechanisms of this evolution and its various stages. On this occasion, it is necessary to emphasise, once again, that those who reject the biological evolution of man have not yet accounted for many of the examples presented by biological events with reference to the biological development of living creatures. Nor have they provided an explanation for those verses in the Qur'an that address the different stages of creation, especially those in Surah al-Sajdah. Any elucidations they might provide should meet the high standards of the Qur'an, its accurate vocabulary, and its precise expressions.

LIFE IN OUTER SPACE

In the previous chapter I mentioned that astronomers have now discovered many planets belonging to many extra-solar systems in our Milky Way. As the estimation is that there are more than 200 billion stars in our galaxy, we would expect them to harbour hundreds of billions of planets. Accordingly, a controversial question has been raised: are any of these planets able to accommodate life? In order to answer this question many researchers have studied and defined the necessary conditions that makes a planet habitable. Data obtained from interstellar dust and space observatories observing the recently discovered exoplanets has been analysed for any indication that they would be suitable for nurturing life. There are a number of crucial factors which constrain the possibility of life on other planets; the most important of these are:

1. Having a solid, stable surface.
2. Having sufficient quantities of liquid water on the planet.
3. Having an atmosphere which contains suitable volumes of gases required for establishing and sustaining life there. The volume, as well as the structure, of the atmosphere should be adequate to protect the surface of the planet from any dangers coming from outer space.
4. To have magnetic shielding with enough strength to protect the planet against the dangers of stellar winds produced by the stars.

Aside from these points noted above, there are, in fact many other requirements for the sustaining of life on extra-solar planets and, as we see in our solar system, certain planets could have accommodated life at some time in their history (like Venus and Mars), but the lack of one or more essential factor meant that the chances of supporting life were no longer feasible. The conditions of establishing and sustaining life in any place of this universe are very difficult to be fulfilled regarding one major perspective: there are many dangers that can eliminate life, whilst the establishment of life, including the more primitive ones, necessitates a lengthy period of time. Several planets, which are similar to Earth in size and mass, have been discovered but there is still a debate as to whether or not they are really qualified for accommodating a flourishing lifeform.

Philosophically, the possibility of life on other planets is an important topic for study in the new *kalām*. In *Daqīq al-kalām*, the physical and biological conditions, along with their implications are assessed, while in *Jalīl al-kalām*, it is the social, moral, and religious questions of such a situation that have to be discussed. A most interesting situation will arise if we were to encounter a creature who is much more advanced than us, for example, creatures who, say, developed millennia before us. Of course, it may be that there is some sort of a natural constraint on the age of advanced civilisations, such that no peoples can progress to a super stage of development before being destroyed. This might well be the case if the example we are experiencing on planet Earth is typical of what is to be found in other places in the universe.

Chapter Fourteen

Research Horizons

Daqīq al-*kalām* brings about a wider scope for research in important issues of natural philosophy. These issues are in close connection to the current problems we find in physics. We have seen how this can be achieved through some applications with examples of fundamental importance laid out in Part III of this book. These examples teach us how powerful the principles of *Daqīq al-kalām* are in providing solutions for those fundamental problems. We have seen the example of utilising the principle of re-creation in offering a solution for the problem of measurement in quantum mechanics in Chapter Eight, how the same principle, along with *kalām* atomism and the description of discrete motion, can provide a solution for the trajectory of a quantum particle in the double slit experiment. We have seen how the principle of contingency and indeterminism can provide a deeper understanding of the anthropic principle through presenting the interplay between fine tuning and the blind laws of nature and, in Chapter Thirteen, I have touched upon using the principle of contingency and indeterminism to seek a possible explanation for the mechanism of biological evolution, and the evolution of the world in general.

In fact, the new *Daqīq al-kalām* carries great potential for offering up solutions to many contentious issues within the natural sciences, specifically in physics and biology. These solutions are urgently needed if there is to be any progress made in these fields. Much of theoretical physics is congested with unnecessary complications brought in by mathematical approaches which have been employed for problems taken out of their physical context. Thus, it might not be surprising, after all, to seek solutions for our problems from natural philosophy.

There are many problems to consider and here are the general guidelines for a set of primary topics that can be put on the study tables of graduates or those experts in these areas. Researchers who deal with such problems need not be philosophers, but they need to understand the principles of *Daqīq* al-*kalām* very well and be specialists in their field of scientific study. Investigating these issues will not necessarily bring about direct solutions for the current scientific problems but may help other experts by injecting new ideas and new paradigms which might enable them to discover new frontiers.

In what follows, I suggest ten problems for study along the new line of thought. These are:

1. The concept of non-existence and the non-existent (the non-being) of the Muʿtazilis, has a connection with the concept of the vacuum in contemporary physics. This school of thought believed that the non-existent is something, contrary to the stance which the Ashʿaris adopted. The Muʿtazilis held that the non-existent are jawāhir (substances) in a vacuum that lacked occupancy (*taḥayyuz*). This means that the non-existent are virtual jawāhir (or virtual particles). Accordingly, a number of questions can be raised. Is it possible to have an agreement between this concept and the concept of the vacuum as offered by the quantum field theory? Is there a useful application for this concept in the relativistic quantum theory of Dirac? What are the implications if we envisage the re-creation of the virtual state to be at work? Can *Daqīq al-kalām* present a clearer vision in this regard?
2. The Mutakallimūn proposed time to be atomised, i.e. being composed of tiny units. Consequently, they hypothesised the presence of a minimum duration of physical time that was indivisible. Congruent with this, space was thought to be discrete and, again, an indivisible minimum distance was envisaged. If this is available, and in the light of the integrity of space and time, can we predict any sort of discreteness in the structure of spacetime? I do not wish to discuss the problem from a physical point of view here, since this topic has been so much deliberated in theoretical physics as to have generated a great deal of confusion. But my questions are: can we find any philosophical realisation within the new *Daqīq al-kalām* that can provide us with a clue for an outset of spacetime discreteness? Can we imagine that the re-creation process is some kind of interplay between space and time? This might well be the case, as long as spacetime is kept invariant; this would explain the laws of conservation in physics.

3. The description of the motion of quantum particles might be very inspiring indeed. Is it possible to extend this description in order to discuss the interactions of particles on the microscopic level, and describe such interactions in terms of motion and re-creation?
4. Can we use the re-creation principle to present a better explanation of the superposition principle in quantum mechanics? Is the superposition of state a real case or is it a mathematical artifact of the theory? Can we provide a more objective solution to Schrödinger's cat paradox? I am sure that there is some scope for that.
5. Another point is relevant to the range of applicability for the laws of quantum mechanics. That is, that some physicists say that the laws of quantum mechanics apply only to microscopic systems, unlike classical mechanical laws which apply to macroscopic systems. The advocates of this view claim that quantum phenomena cannot be seen at the macroscopic level. However, equations show that quantum mechanics applies to all bodies and states in the universe, whether they are microscopic or macroscopic. True, we cannot see quantum effects on the macroscopic level since these are minute effects in such a vast scale. The proposed scheme of re-creation resolves this question. But can we find macroscopic quantum states? How can we understand the relationship between quantum coherence and the existence of macroscopic quantum states in the light of the process of continued re-creation?
6. In Chapter Thirteen I touched upon the co-operative relationship between re-creation and contingency, from which evolution can result. I have also mentioned the possibility of adaptive mutation. Can we analyse these matters further, in order to gain a better understanding of biological evolution?
7. In *Tahāfut al-falāsifah* (*The Incoherence of the Philosophers*), al-Ghazālī presents an elegant dialogue about the necessity of the *particularizer*, where he discusses the motivations for the symmetry-breaking of a system. Can we extend this discussion and take it as an argument for establishing a rule or a principle concerning spontaneous changes that may take place in the physical world? Alternatively, can we argue against al-Ghazālī's argument using the known spontaneous symmetry-breaking in physics?
8. The evidence of creation, or what is today called intelligent design, is an additional theme of paramount importance. Intelligent design is a recent US school of thought which believes that the world is designed in a brilliant way in order to accommodate a conscious and intelligent life system. Some atheists claim that intelligent design is no more than a disguised form of creationism. Is it possible to reconsider the design argument in traditional *kalām* and employ it to construct a consistent view that is more scientific than the intelligent design proposal? Can we use the argument from the principle of least action more efficiently here?

9. The principle of re-creation is very appealing, and I have utilised it in several problems already. However, it is not clear whether, if we have an aggregate of particles, re-creation can be considered for the whole system or for each of its elements only. A detailed discussion is necessary in order to establish a more consistent viewpoint.
10. In his book, *Tahāfut al-falāsifah* (The Incoherence of the Philosophers), Al-Ghazālī, the Muslim theologian, jurist, and philosopher, pointed out that the end of the world and post-eternity has no religious evidence. Yet, this contradicts what we read in Surah Al-Anbia', verse (104) which says, "The day when We roll up heaven like the rolling up of the scroll of writings. As We began the first creation, We shall reproduce it. A promise (binding) on Us. We shall bring it about." Also, in Surah Az-Zumar, verse (67) we read a similar thing about the folding of the Heavens. The question is how can we accurately interpret what al-Ghazālī has written and would be possible that he may have overlooked something?

Briefly, these are ten major problems, amongst others, offered here for research and investigation from the view of the new *Daqīq* al-*kalām*. This, in turn, may establish an advanced philosophy of natural sciences which is not only biased, from an Islamic perspective, but could also be open to the other cultures, civilisations and current scientific trends.

Epilogue

The Islamic vision presented in this book has been aimed at a better understanding of the world. Also, basic pillars have been disclosed for a new curriculum which establishes a strong foundation for original Arab and Islamic thought that widens the horizons of research at different levels. The current approach is not only characterised by philosophical and intellectual capabilities, but it is, hopefully, calculated to reinforce the empirical and scientific character of the approach as well. As was indicated in the heart of the discussions regarding principles and applications, there is room for development of some ideas trying to fuse the original *kalām* principle within the modern context. An example of this is the principle of re-creation which seems to be capable of accommodating several phenomena in quantum physics as well as in biology.

Science can enhance our understanding of religion and even help us explain truly revealed religious scriptures. This is where *kalām* and science can collaborate in order to establish a new vision of the world. I feel that the scheme of thought given in the new *kalām* can play a great role in advancing the understanding of religion and science. I believe that modern science can elevate our understanding of religion and that religion can help provide us with ethical directions for dealing with our scientific knowledge. Modern science can provide us with a greater vision with respect to the facts about the world, while religion can enlighten our vision and guide us to a higher comprehension of the world and our destiny. The lesson that *kalām* teaches us is to have self-awareness of the ignorance that can lead to violence, hatred and calamity, and it teaches the rest of the world the need for more humility and better understanding of the goals of a civilised life.

Being an amalgam of science and belief, *kalām* tells us that we should be able to comprehend the world in a better light, having understood that we humans are the result of a biological evolution that culminated in the receiving of spiritual breath, which transformed us from the state of an animal to the state of human. With this in mind, perhaps we ought to be 'real' humans by continuing our evolution on the spiritual level. Obviously, we can always draw back on our animal qualities and be abducted by our animal heritage in a retrograde step away from the elevation we acquired through receiving the spiritual breath, but that would be degrading ourselves. Highly civilised societies should understand the needs of under-developed societies, and should help them achieve a better status through education, and economic, and social welfare. Otherwise, the materialism of advanced nations possesses no more value than the power possessed by any beast; meaningless and unworthy of respect.

Being confident of this approach, we should realise that we have been bestowed with a considerable mental repertoire as well as the majestic principles inspired by faith. Implementing the mental route should be accompanied by a spiritual guidance, a guidance that need not necessarily be derived from a religious cleric. In other words, the visions of the lesson should be in conformity with the teachings of the Divine message, provided they do not impose a dictatorial power. These poles act to guide our actions and give us the ability to re-adjust our spiritual orientation whenever we are perplexed, or in danger of poor judgement, which may lead to deviation, fanaticism, prejudice, or factionalism. The right approach is fulfilled when we attain an intellectual comprehension enlightened by faith.

The new *kalām* is an ambitious project that aims to transform Muslims and non-Muslims alike with a new insight into God, Man and the world. It also acknowledges each of these with a more advanced and wider vision, which takes lessons from the history and status of man, as a conscious, cognitive entity, into consideration. Such a vision should utilise the mental qualities we have acquired to elevate our understanding of our existence and destiny.

Bibliography

ʿAbd al-Jabbār (1965). Qāḍī Abū al-Ḥasan. *Al-Muḥīṭ bi al-taklīf.* Edited by Omar Sayed Azmi. Cairo: al-Dār al-Miṣrīyya li al-Ta'līf wa al-Tarjam.

ʿAbd al-Jabbār (n.d), Al-Mughnī fi abwab al-ʿadl wa al-tawḥīd, Cairo: Dar al-Maʿrif.

Adamson, P. (2007). *Al-Kindi.* New York: Oxford University Press.

___________ P. (2002). "Abu Maʿshar, al-Kindi and the Philosophical Defense of Astrology." *Recherches de philosophie et théologie médiévales* 69, 245–70.

al-Alūsī, Husam (1965)., The problem of creation in Islamic Thought Qur'an, Ḥadīth, Commentaries and *kalām*, Ph.D. dissertation, Cambridge University.

al-Aṣfahānī, al-Rāghib (1412 A.H.). *al-Mufradāt fi Gharīb al-Qur'ān*, Safwān Dawūdī (ed.) Damascus: Dar al-Qalam.

Al-Ash ʿarī , Abū al-Ḥasan (1980). *Maqālāt al-iIslāmiyyīn wa ikhtilāf al-muṣallīn.* H. Ritter (ed.) Fesbaden: Franz Steire, (3rd ed.).

Al-Baghdadi, (1977). Abdul Qahir, Kitab al-Farq bayn al-Firaq, Beirut: Dar al- afaq al-Jadida.

al-Bāqillānī, (1987). Abū Bakr. *Kitāb Tamhīd al-awā'il wa talkhīṣ al-dalā'il.* Edited by ʿImād al-Dīn Aḥmad Ḥaydar. Beirut: Mu'assasat al-Kutub al-Thaqāfiyya.

al-Bazdawī, Abal-Yusr, *Uṣul al-Dīn*, (1963). Hans Peterlis (ed.), Cairo: Dar Iḥya' al-Kutub al-ʿarabiya.

al-Ghazālī, Abū Ḥāmid (1911). *Mi'yār al-'ilm fi fan al-manṭiq.* Cairo: Kirdistan al-ʿIlmiyyat.

_____ (1990). *On Divine Predicates and their Property*. Edited and translated by ʿAbd al-Raḥmān Abū Zayd. Lahore: Sh. Muhammad Ashraf.

_____ (1980). *Deliverance from Error (al-Munqidh min al-Dalāl)*. Translated by Richard J. McCarthy, S.J. Boston: Twayne.

_____ (2000). *The Incoherence of the Philosophers*. Michael E. Marmura (ed. & trans.). Provo, UT: Brigham Young University Press.

_____ (2004) *Iḥyāʾ Ulūm al-Dīn*, Beirut: Dar al-Marifa.

_____ (2013). *Moderation in Belief*, Aladdin M. Yaqub (trans.), Chicago: Chicago University Press.

al-Iraqi, M. A. (1973). al-Tajdīd fi al-Mazahib al-*Kalām*iyyah wa al-Falsafiyyah, Cairo: Dar al-Maʾarif, , 70.

al-Jaḥiẓ (1963). *Rasaʾil al-Jaḥiẓ*, ʿAbdul-Salam Harūn (ed.), Muʾassasat al-Khānjī.

_____ (2004). Kitab al-Ḥaywan, Beirut: Dar Al-Kutub al-ʿilmiyya.

al-Jurjānī, Alī ibn Muḥammad (1987). *Kitāb al-Taʾrīfāt*. Beirut: Maktabāt Lubnān.

al-Jūwaynī, Abū al-Maʿālī (1969). *Kitāb al-Shāmil fī Uṣūl al-Dīn*. ʿAlī Sāmī Nashshār (ed.). Alexandria: Munshaʾat al-Maʿārif.

al-Khayyāṭ, Abū al-Ḥusayn (1957). *Kitāb al-intiṣār*. Edited by H.S. Nyberg. by Albert N. Nader (Trans.). Beirut: Les Lettres Orientales.

al-Kindī, Yaʿqūb ibn Isḥāq (1953). *Rasāʾil al-Kindī al-falsafiyya*, 2 vols. M. Abū Rīda (ed.). Cairo: Dār al-Fikr al-ʿArabī,

al-Naysābūrī (1979). *Kitab al-Masaʾil fī al-Khilaf*, M.Ziyāda and R. Sayīd (eds.), Beirut: Inmaʾ al-ʿarabī Institute.

Alpher, R. A., Bethe, H., and Gamow, G. (1948). "The origin of chemical elements", Physical Review, 73(7), 803.

al-Rāzī, Zayn al-Dīn (1999). *Mukhtar al-Sihaḥ*, Yousef Shykh Muḥammad (ed.), Beirut: al-ʿAṣriyah printing shop,.

al-Shahristānī, Muḥammad ibn ʿAbd al-Karīm (1934). *Nihāyat al-Iqdām fī ʾilm al-kalām*, Alfred Jeyom (Trns.), Oxford.

_____ (1968). *Kitāb al-milal wa al-niḥal*. Cairo: Muʾassasat al-Ḥalabī.

al-Sīrāfī, Abū Saʿīd. *al-Ghunyah fī Usul al-Dīn*. Imad al-Dīn Haydar (ed.),

Cultual Services Foundation, Beirut, 1987.

Altaie (1978). "Bose-Einstein condensation in an Einstein universe", J. Phys. A (Math. and Gen.) 11, 1603-1620.

Altaie, M. B., and Dowker, J. S., (1978). "Spinor Fields in an Einstein Universe: Finite temperature corrections", Physical Review D18, 3557.

Altaie, M. B. and Malkawie, E (2000). "Bose-Einstein condensation of spin-1 field in an Einstein universe", J. Phys. A (Math. and Gen.) 33, 7093-70102.

Altaie, M. B., Malkawie, A. and Sabbarīnī. M. "Causality in Islamic *Kalām* and in Modern Physics." *Jordanian Journal of Islamic Studies* 8, 2A (2012). Arabic

Altaie, M. B. and al-Zu'bī and M. K. (2008) "The Concept of Heaven and Heavens in the Qur'ān and Modern Astronomy", *Jordanian Journal of Islamic Studies* 4, (3): 223–49.

Altaie, M. B. and al-Ahmad, U (2011). "A Non-singular Universe with Vacuum Energy." *International Journal of Theoretical Physics* 50: 3521–8.

Altaie, M. B., and Setari, M. R. (2003)."Finite-temperature scalar fields and the cosmological constant in an Einstein universe", Physical Review D67, 044018. ________ (1994). "The Scientific Value of *Daqīq al-Kalām.*" *Journal of Islamic Thought and Scientific Creativity* 4 (2): 7–18.

________ (2003). "Back-reaction of the neutrino field in an Einstein universe", Class. and Quantum Gravity 20, 331-340,

________ (2003). "A regional upper limit for particle creation by black holes", Hadronic Journal, 26, 779-794.

________ (2005). "Atomism According to Mutakallimūn." *Etudes Orientales* 23/24, 49–90.

________ (2010). "Re-Creation: A Possible Interpretation of Quantum Indeterminism." In *Matter and Meaning*, edited by Michel Fuller, 21–37. Newcastle Upon Tyne: Cambridge Scholar Publishing.

________ (2015). "The Impact of Philosophical Approach on Late *Kalām*: Analysis of Ibn Khaldūn's Criticism" (conference paper, On Ottoman *Kalām*, Istanbul 29 Mayes University, December 2015).

Aristotle (1929). *Physics*. Wicksteed, P.H. and Cornford, F. (Trans.) Cambridge, MA: Harvard University Press.

Arnaldez, Roger. "Ibn Ḥazm." Islamic Philosophy Online. Translated by Miriam Rosen. Accessed 27 June 2015. <http://www.muslimphilosophy.com/ Ḥazm/ibnḤazm.htm>.

Aspect, Alain (2007). "Quantum Mechanics: To Be or Not to Be Local." *Nature* 446: 866–7.

Bach, R. et al., (2013). Controlled double slit electron diffraction, *New Physics Journal*, 15, 033018.

Baggot, Jim (2013). *Farewell to Reality: How Modern Physics Has Betrayed the Search for Scientific Truth*, London: Pegasus Books.

Barbour, Ian. (1998). *Religion and Science: Historical and Contemporary Issues.* London: SCM Press.

Barrow, J. and Tipler. F. J. (1982). *The Anthropic Cosmological Principle.* Oxford:, Oxford University Press.

Belinfante, Frederik Jozef (1973). *A Survey of Hidden Variables Theories.* Oxford: Pergamon Press.

Bell, J. S. (1988) *Speakable and Unspeakable in Quantum Mechanics: Collected Papers on Quantum Philosophy.* Cambridge: Cambridge University Press,.

________. (1964) "On the Einstein-Podolsky-Rosen Paradox." *Physics* 1 195–200.

Bohm, D. (1951). *Quantum Theory.* New York: Prentice-Hall.

__________ (1952) "A Suggested Interpretation of the Quantum Theory in Terms of 'Hidden' Variables, I and II." *Physical Review* 84 166–79.

Bohm, D. and B. J. Hiley, (1993). *The Undivided Universe: An Ontological interpretation of quantum Theory*, London & New York: Routledge.

Born M., Heisenberg, W. and Jordan. P. (1926) "Zur Quantenmechnik II." *Zeitschrift für Physik* 35 557–615.

Born, M. (1971) *The Born–Einstein Letters,* 1916–1955: *Friendship, Politics and Physics in Uncertain Times.* New York: Walter and Co.; London: Macmillan.

Brandon C. (1974). in Confrontation of cosmological theories with observation, M.S. Longair (ed.), Dordrecht: Reidel.

Cairns, J., Overbaugh, J., & Miller, S. (1988). The origin of mutants. Nature, 335(6186), 142-145.

Carroll, S. (2005). "Why (Almost All) Cosmologists Are Atheists." *Faith and Philosophy 22*, no. (5) 622–35.

Carruthers, P. and Boucher J. (eds.) (1998) Language and Thought: Interdisciplinary Themes. Cambridge: Cambridge University Press.

Cartwright, N. (1983). *How the Laws of Physics Lie.* Oxford: Clarendon Press; New York: Oxford University Press.

___________ . (2005). "No God; No Laws." In Dio, la Natura e la Legge: God and the Laws of Nature, edited by E. Sindoni and S. Moriggi, 183–90. Milan: Angelicum-Mondo X.

Clark, Andy (1997). *Being There: Putting Brain, Body, and World Together Again.* Oxford: Oxford University Press.

__________. (1998). "Magic Words: How Language Augments Human Computation." In *Language and* Thought: Interdisciplinary Themes. Peter Carruthers and Jill Boucher, 162–83. Cambridge: Cambridge University Press.

_________ (2006). "Language, Embodiment, and the Cognitive Niche." *Trends in Cognitive Sciences* 10, no. (8): 370–4.

Cohen, S. Marc Patricia Curd, Charles David, and Chanel Reeve, eds. (2005). *Readings in Ancient Greek Philosophy: From Thales to Aristotle.* 3rd ed. Indianapolis, IN: Hackett.

Craig, William Lane (1979). *The Kalām Cosmological Argument.* London and Basing-stoke: Macmillan Press.

Darādka, Munīr. (2009) "A Collapsing Flat Universe." MSc thesis, Yarmouk University.

Davies, P. (1982). *The Accidental Universe.* Cambridge: Cambridge University Press.

___________ (1990). *God and the New Physics,* London: Penguin Books.

___________ (1992). *The Mind of God,* London: Simon and Schuster.

___________ (2004). "Multiverse Cosmological Models", *Modern Physics Letters .A* vol. 19, no. 10, pp.727-744.

Dawkins, Richard. The Blind Watch Maker. Penguin Books, 2006.

Dennett, Daniel C. (1991). *Consciousness Explained.* New York: Little Brown & Co.

_________ (1993) "Learning and Labeling." *Mind and Language* 8, no. (4):

540–7.

________ (1995). *Darwin's Dangerous Idea: Evolution and the Meanings of Life*. New York: Simon and Schuster.

DeWitt, B. S., and Graham N. (eds.) (1973). *The Many-Worlds Interpretation of Quantum Mechanics: A Fundamental Exposition by Hugh Everett, III, with Papers by J.A. Wheeler, B.S. DeWitt, L. N. Cooper and D. Van Vechten, and N. Graham*. Princeton, NJ: Princeton University Press.

Dhanani, Alnoor (1994). *The Physical Theory of Kalām: Atoms, Space, and Void in Basrian Muʿtazilī Cosmology*, Leiden: Brill.

Dhanani, Alnoor (2015). "The Impact of Ibn Sīnā's Critique of Atomism on Subsequent *Kalām* Discussions of Atomism", *Arabic Sciences and Philosophy*, vol.25: 79-104.

Dicke, Robert H., et al. "Cosmic Black-Body Radiation." The Astrophysical Journal 142 (1965): 414-419.

Dirac, Paul. (1928) "The Quantum Theory of the Electron." *Proceedings of the Royal Society A* 117: 610–24.

Donald, M. J. (2001). "A review of Johnjoe McFadden's book 'Quantum Evolution.' arXiv preprint quant-ph/0101019

Einstein (1936). "Physics and Reality." *Journal of the Franklin Institute* 221, no. 3: 349–82

________ (1987). *Physics and Reality*. New York: Philosophical Library.

Einstein, Albert, B. Podolsky, and N. Rosen. "Can Quantum-Mechanical Description of Physical Reality Be Considered Complete?" *Physical Review* 47 (1935): 777–80.

________. "Physics and Reality." In *Ideas and Opinions*. Sonja Bargmann (Trans.). New York: Bonanza, 1954.

Ellis, George F.R., Ulrich Kirchner, William R. Stoeger, Multiverses and Physical Cosmology, Mon.Not.Roy.Astron.Soc. 347 (2004) 921-936.

__________. (2011). ""Does the Multiverse Really Exist?"", *Scientific American*, 38-43.

Everett III, Hugh (1957). "Relative State Formulation of Quantum Mechanics." *Reviews of Modern Physics* 29: 454–62.

Feynman, R, P., Leighton, R.B., and Matthew, S. (1965). *The Feynman Lectures on Physics*, vol. 1, Reading, MA: Addison-Wesley.

Feynman, R. (1985). *The Character of Physical Law*. Cambridge, MA: MIT Press.

Gabriele V. (2004). "The Myth of the Beginning of Time", *Scientific American*.

Garber, Daniel (2014) "God, Laws, and the Order of Nature: Descartes and Leibniz, Hobbes and Spinoza." In *The Divine Order, the Human Order, and the Order of Nature: Historical Perspectives*, edited by Eric Watkins, 45–66. New York: Oxford University Press.

Gehrels, Neil, Luigi Piro, and Peter JT Leonard, (2002). "The Brightest Explosions in the Universe." *Scientific American* 287, no. 6: 84-91

Gilbert G., Aspect A., and Fabre C. (2010). *Introduction to Quantum Optics: From the Semi-Classical Approach to Quantized Light*, New York: Cambridge University Press.

Golshani, M. (2012). "Quantum Theory, Causality and Islamic Thought." In *The Routledge Companion to Science and Religion*, by James W. Haag, Gregory R. Peterson, and Michael L. Spezio (ed.), 179–90, London: Routledge.

Graham, Neill (1970). "The Everett Interpretation of Quantum Mechanics." PhD thesis, University of North Carolina at Chapel Hill.

Griffel, Frank (2009). *Al-Ghazālī's Philosophical Theology*. Oxford and New York: Oxford University Press.

Grünbaum, (1989). "The Pseudo-Problem of Creation in Physical Cosmology." *Philosophy of Science* 56 (3): 373–94.

_____________ (1991). "Creation as a Pseudo-Explanation in Current Physical Cosmology", *Erkenntinis* 35: 233–54.

Guessom, N. (2011). *Islam's Quantum Question: Reconciling Muslim Tradition and Modern Science*. London: I. B. Tauris.

Hartle, J. B., and Hawking S. W. (1983). "Wave Function of the Universe." *Physical Review D* 28 (12): 2960.

Hawking, S. (1975). "Particle Creation by Black holes", Comm. Math. Phys. 43, 199.

___________. "Black Holes and the Information Paradox." Lecture at GR17: 17th International Conference on General Relativity and Gravitation, Dublin, Ireland, 18–24 July.

____________. (1988). *A Brief History of Time*. New York: Bantam.

____________. (August 2013) "Information Preservation and Weather Forecasting for Black Holes." Lecture at Fuzz or Fire Workshop, The Kalvi Institute for Theoretical Physics, Santa Barbara, CA,. Preprint available at arXive: 1401.5761v1, accessed 22 June.

Hawking, S. and Ellis, George F. R., (1973). *The Large Scale Structure of Space-Time.* Cambridge and New York: Cambridge University Press.

Hawking, Stephen, and Mlodinow, L. (2010). *The Grand Design.* New York: Bantam Books,

Heisenberg, W. (1925). "Über quantentheoretische Umdeutung kinematischer und mechanischer Beziehungen." *Zeitschrift für Physik* 33 879–93.

__________ (1926). Math. Ann. 95.

__________ (1999). *Physics and Philosophy: The Revolution in Modern Science.* New York: Prometheus.

Hoefer, Carl (2008). "Causal Determinism." In The Stanford Encyclopedia of Philosophy, edited by Edward N.Zalta. Stanford University, <http://plato.stanford.edu/archives/ win2008/ entries/determinism-causal/>.

Holton, G. J. (1980). "Einstein's Scientific Program: The Formative Years." In *Some Strangeness in the Proportion: A Centennial Symposium to Celebrate the Achievements of Albert Einstein* edited by Harry Woolf, 57–76· Reading, MA: Addison-Wesley Pub. Co.

Hoyle, F. Burbidge, E. Margaret, et al. (1957) "Synthesis of the elements in stars." Reviews of modern physics 29.4: 547.

Hubble, Erwin (1943). *American Scientist* 43(2).

Ibn Faris, Asim (1979). *Mujam Maqayīs al-Lugha*, edited by Abdul Salam Harūn, Dar al-Fikr, Beruit.

Ibn Ḥanbal, Ahmad. The Musnad, Risāla Edition, vol. 29, 385

Ibn Ḥazm, ʿAlī ibn Aḥmad (1964). *Kitāb al-fiṣal fi al-milal wa al-ahw'wa al-niḥal.* Egypt: Mu'assasat al-Khānjī.

Ibn Khaldūn (1958). *The Muqaddimah: An Introduction to History.* Translated by Franz Rosenthal. New York: Pantheon Books.

Ibn Manẓūr (1993). *Lisan al 'Arab.* Beirut: Dar Ṣadir.

Ibn Matawayh (1975). al-Ḥasan ibn Aḥmad. *Kitāb al-tadhkira fi aḥkām al-jawāhir wa al-a'rḍ.* Sami Lutf (ed.), Cairo: al-Dār al-Thaqāfa.

Ibn Rushd (1921). *The Philosophy and Theology of Averroes* [*Faṣl al-maqāl*].

Mohammed Jamil al-Rahman (trans.). Baroda: A.G. Widgery.

___________. (1954). *Tahaāfut al-Tahaāfut (The Incoherence of the Incoherence).* Simon Van Den Bergh (trans. & ed.). London: Trustees of the E.J.W. Gibb Memorial.

___________. (1959) al-Kashf ʿan Manahej al-Adella fi ʿ*A*qaid al-Milla, Arabic Text, edited by George F. Haurani, Leyde (E. J. Brill).

Itano, W. H. and Heinzen, D.J., Bollinger, J. J. and Wineland, D. J. (1990). Phys. Rev. A 41: 2295.

İzmirlī, İsmāʿīl Haqqī (1915). *Islam'da felsefe: Yeni 'ilm-i kalām.* Sebilurrashād XIV/344:43.

James A. Sanders, (1984) *Canon and Community: A Guide to Canonical Criticism*, Philadelphia: Fortress Press.

Jammer, M. (1974). *The Philosophy of Quantum Mechanics: The Interpretations of Quantum Mechanics in Historical Perspective.* New York: Wiley-Interscience.

____________. (1969). *Concepts of Space: The History of the Concepts of Space in Physics.* (2nd ed.), Cambridge, MA: Harvard University Press.

Jauch, J. M. (1973). "The Problem of Measurement in Quantum Mechanics." In *The Physicist's Conception of Nature*, Jagdish Mehra (ed.), 684–6. Boston: D. Reidel Publishing Co.,

Kamali, Mohammad Hashim. (2015). "Causality and Divine Action: The Islamic Perspective." Ghazali.org. Accessed 26 Junehttp://www.ghazali.org/ articles/kamali.htm.

Khan, Syed Ahmad. "Lecture on Islam." In *Sayyid Ahmad Khan. A Reinterpretation of Muslim Theology*, Christian W. Troll (ed.). New Delhi: Vikas Publ. House, 1978.

King, J. L. and Thomas H. Jukes (1969) "Non-Darwinian Evolution" Science 164 (3881): 788–798.

Krauss, L. M. (2012). *A Universe from Nothing: Why There is something rather than nothing.* New York: Free Press.

Lameter, C. (2004). "Divine Action in the Framework of Scientific Knowledge." PhD thesis, Fuller Theological Seminary, Newark, CA,

Laplace, P. (1951). *A Philosophical Essay on Probabilities.* lated by F. W. Truscott and F. L. Emory (Trans.), New York: Dover,

Linde, A. (November 1994). "The Self-Reproducing Inflationary Universe", *Scientific American*,

Linde, A. D. (1983). Physics Letters 129B, 177.

____________. (1990). *Particle Physics and Inflationary Cosmology*, Harwood Academic Publishers, Chur, Switzerland.

Maimonides, M. (1904). *The Guide for the Perplexed.* lated by Michael Friedländer. (Trans.) 2nd ed. London: Routledge & Kegan Paul Ltd.

Marc Cohen, S., Patricia Curd, Charles David, and Chanel Reeve (2005) (eds.), Readings in Ancient Greek Philosophy: From Thales to Aristotle, 3rd ed. Indianapolis, IN: Hackett.

Marmura, M. E. (1984) (ed.) *Islamic Theology and Philosophy.* New York: State University of New York Press,

McFadden, J. and Al-Khalili, J. (1999). "A quantum Mechanical Model of Adaptive Mutation", BioSystems 50 203–211.

McFadden, J., and Al-Khalili, J. (2001). "Comment on Book Review of Quantum Evolution'(Johnjoe McFadden) by Mathew J. Donald." arXiv preprint quant-ph/0110083.

Meadows, V. S. (2008), *Planetary Environmental Signatures for Habitability and Life,* in Mason, J. (Editor), Exoplanets: detection, Formation, Properties, Habitability, Springer.

Namiki, Mikio, and Saverio Pascazio (1993). "Quantum Theory of Measurement Based on the Many-Hilbert-Space Approach." *Physics Report* 232 (6) 301–411.

Newton, Isaac (2010). *The Principia: Mathematical Principles of Natural Philosophy.* Snowball Publishing,

Nuʿmānī, Shiblī (1929). *'Ilm al-kalām.* Karachi.

Ossendrijver, Mathieu. (January 2016). "Ancient Babylonian astronomers calculated Jupiter's position from the area under a time-velocity graph", Science, Vol. 351 (6272),

Ostlie, D.A. and Carroll, B. W. (2007). *An Introduction to Modern Stellar Astrophysics.* Addison Wesley, San Francisco.

Özer, M., and Taha M. O. (1987). "A Model of the Universe Free of Cosmological Problems." *Nuclear Physics B* 287 776–96.

Özervarlı, M. Sait (1999) "Attempts to Revitalize *Kalām* in the Late 19th and Early 20th Centuries." *The Muslim World* 89 (1), 90–105.

Penzias, Arno A., and Robert Woodrow Wilson. (1965). "A measurement of excess antenna temperature at 4080 MHz" The Astrophysical Journal 142: 419-421.

Petrosky, T. Tasaki, S. and Prigogine, I. (1990). Phys. Lett. A 151: 109.

Pines, S. (1997). *Studies in Islamic Atomism.* Translated by Michael Schwarz. Edited by Tzvi Langermann. Jerusalem: The Magnes Press.

Polkinghorne, John (1990). "God's Action in the World." J. K. Russell Fellowship Lecture, Pacific School of Religion Chapel, Berkeley, CA.

____________. (1993) *Reason and Reality: The Relationship Between Science and Theology*. London: SPCK.

____________. (1995) "The Metaphysics of Divine Action." In *Chaos and Complexity: Scientific Perspectives on Divine Action*, vol. 2. Robert John Russell, Nancey Murphy, and Arthur Peacocke (eds.), 147–56. Vatican City State: Vatican Observatory; Berkeley, CA: Center for Theology and the Natural Sciences.

Povh, B., Rith, K., Scholz, C., Zetsche, F. (2002). Particles and Nuclei: An Introduction to the Physical Concepts. Berlin: Springer-Verlag, 73.

Pretzlerzil, Otto (1931), "Die früuhislamische Atomlehre." *Der Islam*, 19, 117–30.

Ragep, F. Jamil (2008). "When did Islamic Science Die (and Who Cares?)." *Viewpoint: The Newsletter of the British Society for History of Science*, no. 85: 1–3.

Rees, M. J (2001), "Concluding Perspective", astro-ph/0101268.

___________ (2001). *Just Six Numbers: The Deep Forces that Shape the Universe* (Basic Books).

___________ (2001). *Our Cosmic Habitat*, Princeton University Press.

Richard J Szabo, *An introduction to string theory and D-brane dynamics*, Imperial College Press; 2 edition, (2011).

Salaris, Maurizio; Cassisi, Santi (2005), *Evolution of stars and stellar populations*, John Wiley and Sons.

Sanders, James A. (1984). *Canon and Community: A Guide to Canonical Criticism* (Philadelphia: Fortress Press.

Schrödinger, E (1926). "Quantisierung als Eigenwertproblem (Erste Mitteilung)." *Annalen der Physik* 79 (4), 361–76.

__________ (1935). "Die Gegenwärtige Situation in der Quantenmechanik [The Present Situation in Quantum Mechanics]." *Naturwissenschaften* 33 (November).

Schulman, L. S. (1994). Physica Scripta 49: 536-542.

__________ (1997). J. Phys. A: Math. Gen. 30: L.293-299.

Silk, J. (1977). "Cosmogony and the magnitude of the dimensionless gravitational coupling constant" Nature, 265, 710-11.

Smith, Q. (1994). "Can everything come to be without cause?" Dialogue 33: 313-323,

________ (1988). "The uncaused beginning of the Universe", Philosophy of science 55: 39-57.

Smirnov, André (1997). "*Causality and Islamic Thought*", published in "A Companion to World Philosophies", ed. E. Deutch and R. Bontekoe, Blackwell publishers, pp.493-503.

Smolin, Lee (2006).*The Trouble with Physics: The Rise of String Theory, the Fall of a Science and What Comes Next.* Boston: Houghton Mifflin.

Staune, J (2007). *Does our existence have a meaning*? (Paris: Presses de la Renaissance).

Swerdlow, N. M. and Neugebauer, O. (1984). *Mathematical Astronomy in Copernicus's 'De revolutionibus.'* New York: Springer-Verlag.

Szabo, Richard J. (2011)., *An introduction to string theory and D-brane dynamics*, Imperial College Press; 2 edition.

Taylor, Edwin F., and Wheeler, J.A. (2000). *Exploring Black Holes.* New York: Addison-Wesley Longman,.

Tegmark, M. (2003) "Parallel Universes", *Scientific American*, 41–8.

_________ (February 2014) "Are Parallel Universes Unscientific Nonsense?" Insider Tips for Criticizing the Multiverse, *Scientific American.*

Versteegh, Kees. (1997). *The Arabic Linguistic Tradition.* Landmarks in Linguistic Thought, vol. 3. London and New York: Routledge,

von Neumann, John (1955). *Mathematical Foundations of Quantum Mechanics.* Translated by Robert T. Beyer. Princeton, NJ: Princeton University Press,

Vygotsky (1978). *Mind in Society*. Cambridge, MA: Harvard University Press.

_______ (1962). *Thought and Language*. Cambridge, MA: MIT Press.

Walter W. R. B. (2003) "Pierre Simon Laplace (1749–1827)", *in A Short Account of the History of Mathematics*, 4th ed. (New York: Dover Publications).

Walzer, R. (1970). "Early Islamic Philosophers." In *The Cambridge History of Late Greek and Early Medieval Philosophy*, edited by A.H. Armstrong, 643–51. Cambridge: Cambridge University Press.

Ward, K. Pascal, Fire: Scientific Faith and Religious Understanding, (2006). Oxford: One World.

Weinberg, S. (1972). *Gravitation and Cosmology*, New York: John Wiley & Sons.

_________ (1979). *The First Three Minutes*, New York: Bantam.

_________ (2001)."The cosmological constant problems." In *Sources and Detection of Dark Matter and Dark Energy in the Universe*, pp. 18-26. Springer, Berlin, Heidelberg,

___________ (2002) In a Debate with John Polkinghorne at the SSQ symposium "Science and the Three Monotheisms: A New Partnership", Granada, Spain, Sept.23–25.

___________ (2008). "Without God." *The New York Review of Books*. 25 September

Wheeler, J. A. (1957). "Assessment of Everett's 'Relative State' Formulation of Quantum Theory." *Reviews of Modern Physics* 29: 463–5.

Wheeler, J. A. and Wojciech H. Z. (1983). *Quantum Theory and Measurement*. Princeton: Princeton University Press.

Wolfson, H. A. (1976). *The Philosophy of the Kalām*. Cambridge, MA: Harvard University Press.

Index

F

G

H

I

R

S

T

U

V